T0250160

Honda TRX300 ATV Owners Workshop Manual

by Alan Ahlstrand and John H Haynes
Member of the Guild of Motoring Writers

Models covered:
TRX300 (Fourtrax 300), 1988 through 2000
TRX300FW (Fourtrax 300 4x4), 1988 and 1990 through 2000

(2125 - 10N2)

**Haynes Group Limited
Haynes North America, Inc.**

www.haynes.com

Acknowledgments

Our thanks to Honda of Milpitas, Milpitas, California, for providing the facilities used for these photographs; to Pete Sirett, service manager, for arranging the facilities and fitting the mechanical work into his shop's busy schedule; and to Steve Van Horn, service technician, for doing the mechanical work and providing valuable technical information.

A book in the Haynes Owners Workshop Manual Series

ISBN-10: 1-56392-439-0

ISBN-13: 978-1-56392-439-2

Library of Congress Control Number: 2001095064

British Library Cataloguing in Publication Data
A catalogue record for this book is available from the British Library

Contents

Introductory pages

About this manual	0-5
Introduction to the Honda TRX300	0-5
Identification numbers	0-6
Buying parts	0-7
General specifications	0-7
Maintenance techniques, tools and working facilities	0-9
Safety first!	0-15
ATV chemicals and lubricants	0-16
Troubleshooting	0-17

Chapter 1

Tune-up and routine maintenance	1-1

Chapter 2

Engine, clutch and transmission	2-1

Chapter 3

Fuel and exhaust systems	3-1

Chapter 4

Ignition system	4-1

Chapter 5

Steering, suspension and final drive	5-1

Chapter 6

Brakes, wheels and tires	6-1

Chapter 7

Frame and bodywork	7-1

Chapter 8

Electrical system	8-1

Chapter 9

Wiring diagrams	9-1

Conversion factors

Index IND-1

Honda TRX300 (1995 4WD model)

About this manual

Its purpose

The purpose of this manual is to help you get the best value from your vehicle. It can do so in several ways. It can help you decide what work must be done, even if you choose to have it done by a dealer service department or a repair shop; it provides information and procedures for routine maintenance and servicing; and it offers diagnostic and repair procedures to follow when trouble occurs.

We hope you use the manual to tackle the work yourself. For many simpler jobs, doing it yourself may be quicker than arranging an appointment to get the vehicle into a shop and making the trips to leave it and pick it up. More importantly, a lot of money can be saved by avoiding the expense the shop must pass on to you to cover its labor and overhead costs. An added benefit is the sense of satisfaction and accomplishment that you feel after doing the job yourself.

Using the manual

The manual is divided into Chapters. Each Chapter is divided into numbered Sections, which are headed in bold type between horizontal lines. Each Section consists of consecutively numbered paragraphs.

At the beginning of each numbered Section you will be referred to any illustrations which apply to the procedures in that Section. The reference numbers used in illustration captions pinpoint the pertinent Section and the Step within that Section. That is, illustration 3.2 means the illustration refers to Section 3 and Step (or paragraph) 2 within that Section.

Procedures, once described in the text, are not normally repeated. When it's necessary to refer to another Chapter, the reference will be given as Chapter and Section number. Cross references given without use of the word "Chapter" apply to Sections and/or paragraphs in the same Chapter. For example, "see Section 8" means in the same Chapter.

References to the left or right side of the vehicle assume you are sitting on the seat, facing forward.

All-terrain vehicle manufacturers continually make changes to specifications and recommendations, and these, when notified, are incorporated into our manuals at the earliest opportunity.

Even though we have prepared this manual with extreme care, neither the publisher nor the author can accept responsibility for any errors in, or omissions from, the information given.

NOTE

A **Note** provides information necessary to properly complete a procedure or information which will make the procedure easier to understand.

CAUTION

A **Caution** provides a special procedure or special steps which must be taken while completing the procedure where the Caution is found. Not heeding a Caution can result in damage to the assembly being worked on.

WARNING

A **Warning** provides a special procedure or special steps which must be taken while completing the procedure where the Warning is found. Not heeding a Warning can result in personal injury.

Introduction to the Honda TRX300

The TRX300 (Fourtrax 300) series are highly successful and popular utility all-terrain vehicles.

The engine on all models is an air-cooled single with an overhead camshaft.

Fuel is delivered to the cylinder by a single constant velocity carburetor.

The front suspension on 1988 through 1992 2wd models uses a coil spring-shock absorber unit and a single control arm on each side of the vehicle. All other models use upper and lower control arms on each side of the vehicle, with a coil spring-shock absorber unit attached to each upper control arm.

The rear suspension on all models uses a single shock absorber and coil spring.

The front brakes use a sealed drum brake at each wheel; the rear uses a single drum brake, mounted on the axle shaft inboard of the wheels. All three brake units are sealed to keep out water and dirt.

Shaft final drive is used at the rear of all models covered in this manual. On 4wd models, power is transmitted to the front wheels by the transmission countershaft to an output gear, from the output gear by a driveshaft to the transfer case, from the transfer case by another driveshaft to the front differential, and from the front differential through driveaxles to the front wheel hubs.

Identification numbers

The frame serial number is stamped into the front of the frame and printed on a label affixed to the frame. The engine number is stamped into the right side of the crankcase. Both of these numbers should be recorded and kept in a safe place so they can be furnished to law enforcement officials in the event of a theft.

The frame serial number, engine serial number and carburetor identification number should also be kept in a handy place (such as with your driver's license) so they are always available when purchasing or ordering parts for your machine.

The models covered by this manual are as follows:

TRX300 (Fourtrax 300), 1988 through 2000
TRX300FW (Fourtrax 4x4), 1988 and 1990 through 2000

Identifying model years

The procedures in this manual identify the bikes by model year. The model year is included in a decal on the frame, but in case the decal is missing or obscured, the following table identifies the initial frame number of each model year.

Year	Initial frame number
1988	
2wd	TE140*JK000018 on
4wd	TE150*JK000018 on
1989 (2wd)	JH3TE140*KK100003 on
1990	
2wd	478TE140*LA200001 on
4wd	478TE150*LA200001 on
1991	
2wd (except New Hampshire)	478TE140*MA300001 on
2wd (New Hampshire)	478TE143*MA300001 on
4wd (except New Hampshire)	478TE150*MA300001 on
4wd (New Hampshire)	478TE153*MA300001 on
1992	
2wd (except New Hampshire)	478TE140*NA300001 on
2wd (New Hampshire)	478TE143*NA300001 on
4wd (except New Hampshire)	478TE150*NA300001 on
4wd (New Hampshire)	478TE153*NA300001 on
1993	
2wd (except New Hampshire)	478TE140*PA300001 on
2wd (New Hampshire)	478TE143*PA300001 on
4wd (except New Hampshire)	478TE150*PA300001 on
4wd (New Hampshire)	478TE153*PA300001 on
1994	
2wd (except New Hampshire)	TE140*RA600001 on
2wd (New Hampshire)	TE143*RA600001 on
4wd (except New Hampshire)	TE153*RA600001 on
4wd (New Hampshire)	TE150*RA600001 on

Year	Initial frame number
1995	
2wd (except New Hampshire)	TE140*SA700001 on
2wd (New Hampshire)	TE143*SA700001 on
4wd (except New Hampshire)	TE150*SA700001 on
4wd (New Hampshire)	TE153*SA700001 on
1996	
2wd (except New Hampshire)	TE140*TA800001
2wd (New Hampshire)	TE143*TA800001
4wd (except New Hampshire)	TE150*TA800001
2wd (New Hampshire)	TE153*TA800001
1997	
2wd (except New Hampshire)	TE140*VA900001
2wd (New Hampshire)	TE143*VA900001
4wd (except New Hampshire)	TE150*VA900001
2wd (New Hampshire)	TE153*VA900001
1998	
2wd (except California)	TE140*WA900001
2wd (California)	TE141*WA900001
4wd (except California)	TE150*WA900001
2wd (California)	TE151*WA900001
1999	
2wd (except California)	TE140*XA000001
2wd (California)	TE141*XA000001
4wd (except California)	TE150*XA000001
2wd (California)	TE151*XA000001
2000	
2wd (except California)	TE140*YA100001
2wd (California)	TE141*YA100001
4wd (except California)	TE150*YA100001
2wd (California)	TE151*YA100001

The frame serial number is located at the front of the frame

The engine serial number is located on the right side of the crankcase behind the cylinder

Buying parts

Once you have found all the identification numbers, record them for reference when buying parts. Since the manufacturers change specifications, parts and vendors (companies that manufacture various components on the machine), providing the ID numbers is the only way to be reasonably sure that you are buying the correct parts.

Whenever possible, take the worn part to the dealer so direct comparison with the new component can be made. Along the trail from the manufacturer to the parts shelf, there are numerous places that the part can end up with the wrong number or be listed incorrectly.

The two places to purchase new parts for your vehicle - the accessory store and the franchised dealer - differ in the type of parts they carry. While dealers can obtain virtually every part for your vehicle, the accessory dealer is usually limited to normal high wear items such as shock absorbers, tune-up parts, various engine gaskets, cables, chains, brake parts, etc. Rarely will an accessory outlet have major suspension components, cylinders, transmission gears, or cases.

Used parts can be obtained for roughly half the price of new ones, but you can't always be sure of what you're getting. Once again, take your worn part to the wrecking yard (breaker) for direct comparison.

Whether buying new, used or rebuilt parts, the best course is to deal directly with someone who specializes in parts for your particular make.

General specifications

Wheelbase
  1988 through 1990 models
    2wd .... 1245 mm (49.0 inches)
    4wd .... 1235 mm (48.6 inches)
  1991 and 1992 models
    2wd .... 1250 mm (49.2 inches)
    4wd .... 1235 mm (48.6 inches)
  1993 and later models
    2wd .... 1239 mm (48.8 inches)
    4wd
      1993 through 1997 .... 1238 mm (48.7 inches)
      1998 and later.. 1235 mm (48.6 inches)
Overall length
  1988 through 1990 models
    2wd .... 1905 mm (75.0 inches)
    4wd .... 1895 mm (74.6 inches)
  1991 and later models (2wd and 4wd)... 1910 mm (75.2 inches)
Overall width
  1988 through 1990 models
    2wd .... 1115 mm (43.9 inches)
    4wd .... 1065 mm (41.9 inches)
  1991 and 1992 models
    2wd .... 1125 mm (44.3 inches)
    4wd .... 1110 mm ((43.7 inches)
  1993 through 1997 models
    2wd .... 1115 mm (43.9 inches)
    4wd .... 1110 mm (43.7 inches)
  1998 and later models... 1157 mm (45.6 inches)

Overall height
    1988 and 1990 models.. 1055 mm (41.5 inches)
    1991 and 1992 models
        2wd .... 1075 mm (42.3 inches)
        4wd .... 1085 mm (42.7 inches)
    1993 through 1997 models
        2wd .... 1088 mm (42.8 inches)
        4wd .... 1085 mm (42.7 inches)
    1998 and later models
        2wd .... 1077 mm (42.4 inches)
        4wd .... 1089 mm (42.9 inches)
Seat height
    1988 through 1990 models
        2wd .... 765 mm (30.1 inches)
        4wd .... 800 mm (31.5 inches)
    1991 and 1992 models
        2wd .... 780 mm (30.7 inches)
        4wd .... 790 mm (31.1 inches)
    1993 and later models
        2wd .... 783 mm (30.8 inches)
        4wd
            1993 through 1997 .... 780 mm (30.7 inches)
            1998 and later... 788 mm (31.0 inches)
Ground clearance
    1988 through 1997 .... 160 mm (6.3 inches)
    1998 and later
        2wd .... 160 mm (6.3 inches)
        4wd .... 172 mm (6.8 inches)
Dry weight
    1988 models
        2wd .... 199 kg (239 lbs)
        4wd .... 215 kg (474 lbs)
    1989 and 1990 models
        2wd .... 211.5 kg (466 lbs)
        4wd .... 230 kg (507 lbs)
    1991 and 1992 models
        2wd .... 216 kg (476 lbs)
        4wd .... 236 kg (520 lbs)
    1993 and later models
        2wd .... 221 kg (487 lbs)
        4wd .... 239 kg (527 lbs)

Maintenance techniques, tools and working facilities

Basic maintenance techniques

There are a number of techniques involved in maintenance and repair that will be referred to throughout this manual. Application of these techniques will enable the amateur mechanic to be more efficient, better organized and capable of performing the various tasks properly, which will ensure that the repair job is thorough and complete.

Fastening systems

Fasteners, basically, are nuts, bolts and screws used to hold two or more parts together. There are a few things to keep in mind when working with fasteners. Almost all of them use a locking device of some type (either a lock washer, locknut, locking tab or thread adhesive). All threaded fasteners should be clean, straight, have undamaged threads and undamaged corners on the hex head where the wrench fits. Develop the habit of replacing all damaged nuts and bolts with new ones.

Rusted nuts and bolts should be treated with a penetrating oil to ease removal and prevent breakage. Some mechanics use turpentine in a spout type oil can, which works quite well. After applying the rust penetrant, let it "work" for a few minutes before trying to loosen the nut or bolt. Badly rusted fasteners may have to be chiseled off or removed with a special nut breaker, available at tool stores.

If a bolt or stud breaks off in an assembly, it can be drilled out and removed with a special tool called an E-Z out (or screw extractor). Most dealer service departments and vehicle repair shops can perform this task, as well as others (such as the repair of threaded holes that have been stripped out).

Flat washers and lock washers, when removed from an assembly, should always be replaced exactly as removed. Replace any damaged washers with new ones. Always use a flat washer between a lock washer and any soft metal surface (such as aluminum), thin sheet metal or plastic. Special locknuts can only be used once or twice before they lose their locking ability and must be replaced.

Tightening sequences and procedures

When threaded fasteners are tightened, they are often tightened to a specific torque value (torque is basically a twisting force). Over-tightening the fastener can weaken it and cause it to break, while under-tightening can cause it to eventually come loose. Each bolt, depending on the material it's made of, the diameter of its shank and the material it is threaded into, has a specific torque value, which is noted in the Specifications. Be sure to follow the torque recommendations closely.

Fasteners laid out in a pattern (i.e. cylinder head bolts, engine case bolts, etc.) must be loosened or tightened in a sequence to avoid warping the component. Initially, the bolts/nuts should go on finger tight only. Next, they should be tightened one full turn each, in a criss-cross or diagonal pattern. After each one has been tightened one full turn, return to the first one tightened and tighten them all one half turn, following the same pattern. Finally, tighten each of them one quarter turn at a time until each fastener has been tightened to the proper torque. To loosen and remove the fasteners the procedure would be reversed.

Disassembly sequence

Component disassembly should be done with care and purpose to help ensure that the parts go back together properly during reassembly. Always keep track of the sequence in which parts are removed. Take note of special characteristics or marks on parts that can be installed more than one way (such as a grooved thrust washer on a shaft). It's a good idea to lay the disassembled parts out on a clean surface in the order that they were removed. It may also be helpful to make sketches or take instant photos of components before removal.

When removing fasteners from a component, keep track of their locations. Sometimes threading a bolt back in a part, or putting the washers and nut back on a stud, can prevent mixups later. If nuts and bolts can't be returned to their original locations, they should be kept in a compartmented box or a series of small boxes. A cupcake or muffin tin is ideal for this purpose, since each cavity can hold the bolts and nuts from a particular area (i.e. engine case bolts, valve cover bolts, engine mount bolts, etc.). A pan of this type is especially helpful when working on assemblies with very small parts (such as the carburetors and the valve train). The cavities can be marked with paint or tape to identify the contents.

Whenever wiring looms, harnesses or connectors are separated, it's a good idea to identify the two halves with numbered pieces of masking tape so they can be easily reconnected.

Gasket sealing surfaces

Throughout any vehicle, gaskets are used to seal the mating surfaces between components and keep lubricants, fluids, vacuum or pressure contained in an assembly.

Many times these gaskets are coated with a liquid or paste type gasket sealing compound before assembly. Age, heat and pressure can sometimes cause the two parts to stick together so tightly that they are very difficult to separate. In most cases, the part can be loosened by striking it with a soft-faced hammer near the mating surfaces. A regular hammer can be used if a block of wood is placed between the hammer and the part. Do not hammer on cast parts or parts that could be easily damaged. With any particularly stubborn part, always recheck to make sure that every fastener has been removed.

Avoid using a screwdriver or bar to pry apart components, as they can easily mar the gasket sealing surfaces of the parts (which must remain smooth). If prying is absolutely necessary, use a piece of wood, but keep in mind that extra clean-up will be necessary if the wood splinters.

After the parts are separated, the old gasket must be carefully scraped off and the gasket surfaces cleaned. Stubborn gasket material can be soaked with a gasket remover (available in aerosol cans) to soften it so it can be easily scraped off. A scraper can be fashioned from a piece of copper tubing by flattening and sharpening one end. Copper is recommended because it is usually softer than the surfaces to be scraped, which reduces the chance of gouging the part. Some gaskets can be removed with a wire brush, but regardless of the method used, the mating surfaces must be left clean and smooth. If for some reason the gasket surface is gouged, then a gasket sealer thick enough to fill scratches will have to be used during reassembly of the components. For most applications, a non-drying (or semi-drying) gasket sealer is best.

Hose removal tips

Hose removal precautions closely parallel gasket removal precautions. Avoid scratching or gouging the surface that the hose mates against or the connection may leak. Because of various chemical reactions, the rubber in hoses can bond itself to the metal spigot that the hose fits over. To remove a hose, first loosen the hose clamps that secure it to the spigot. Then, with slip joint pliers, grab the hose at the clamp and rotate it around the spigot. Work it back and forth until it is completely free, then pull it off (silicone or other lubricants will ease removal if they can be applied between the hose and the outside of the spigot). Apply the same lubricant to the inside of the hose and the outside of the spigot to simplify installation.

Spark plug gap adjusting tool

Feeler gauge set

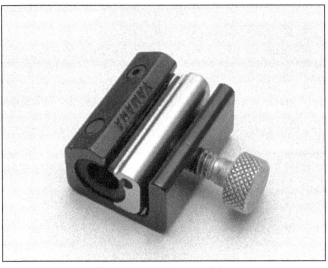

Control cable pressure luber

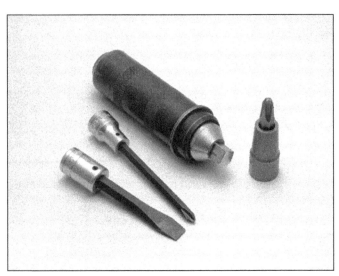

Hand impact screwdriver and bits

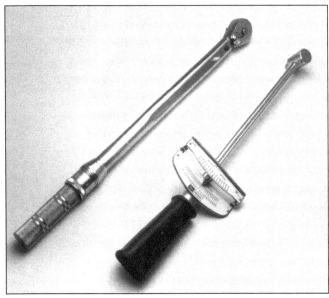

Torque wrenches (left - click type; right, beam type)

If a hose clamp is broken or damaged, do not reuse it. Also, do not reuse hoses that are cracked, split or torn.

Tools

A selection of good tools is a basic requirement for anyone who plans to maintain and repair a vehicle. For the owner who has few tools, if any, the initial investment might seem high, but when compared to the spiraling costs of routine maintenance and repair, it is a wise one.

To help the owner decide which tools are needed to perform the tasks detailed in this manual, the following tool lists are offered: Maintenance and minor repair, Repair and overhaul and Special. The newcomer to practical mechanics should start off with the Maintenance and minor repair tool kit, which is adequate for the simpler jobs. Then, as confidence and experience grow, the owner can tackle more difficult tasks, buying additional tools as they are needed. Eventually the basic kit will be built into the Repair and overhaul tool set. Over a period of time, the experienced do-it-yourselfer will assemble a tool set complete enough for most repair and overhaul procedures and will add tools from the Special category when it is felt that the expense is justified by the frequency of use.

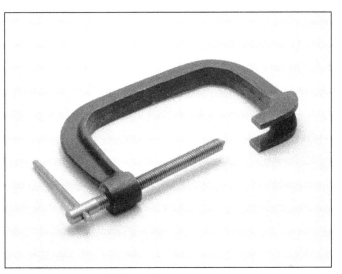

Snap-ring pliers (top - external; bottom - internal)

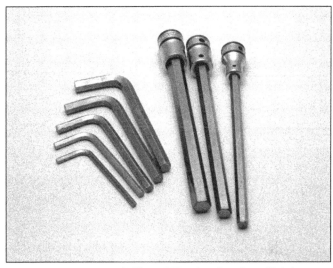

Allen wrenches (left), and Allen head sockets (right)

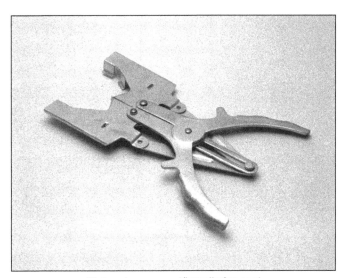

Valve spring compressor

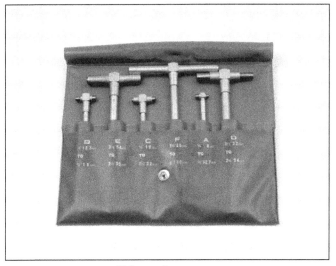

Piston ring removal/installation tool

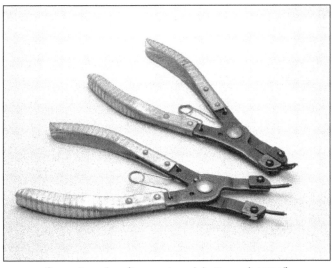

Piston pin puller

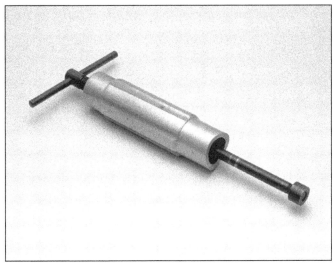

Telescoping gauges

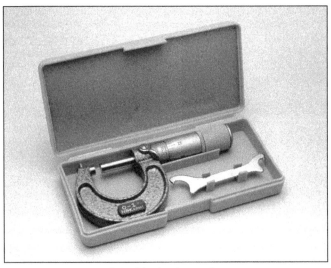

0-to-1 inch micrometer

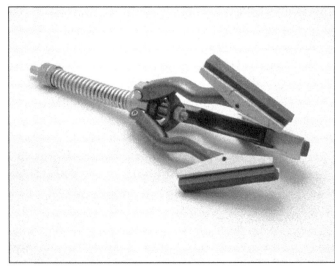

Cylinder surfacing hone

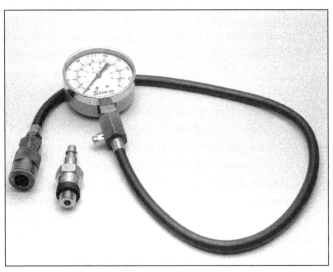

Cylinder compression gauge

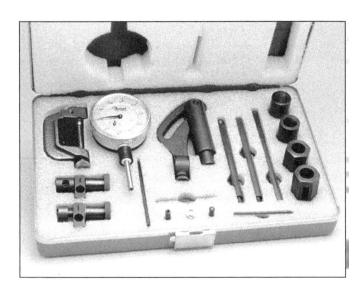

Dial indicator set

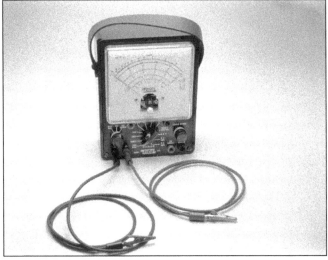

Multimeter (volt/ohm/ammeter)

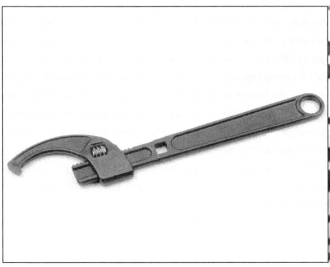

Adjustable spanner

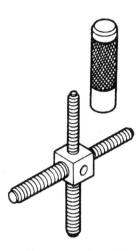

Alternator rotor puller

Maintenance and minor repair tool kit

The tools in this list should be considered the minimum required for performance of routine maintenance, servicing and minor repair work. We recommend the purchase of combination wrenches (box end and open end combined in one wrench); while more expensive than open-ended ones, they offer the advantages of both types of wrench.

Combination wrench set (6 mm to 22 mm)
Adjustable wrench - 8 in
Spark plug socket (with rubber insert)
Spark plug gap adjusting tool
Feeler gauge set
Standard screwdriver (5/16 in x 6 in)
Phillips screwdriver (No. 2 x 6 in)
Allen (hex) wrench set (4 mm to 12 mm)
Combination (slip-joint) pliers - 6 in
Hacksaw and assortment of blades
Tire pressure gauge
Control cable pressure luber
Grease gun
Oil can
Fine emery cloth
Wire brush
Hand impact screwdriver and bits
Funnel (medium size)
Safety goggles
Drain pan
Work light with extension cord

Repair and overhaul tool set

These tools are essential for anyone who plans to perform major repairs and are intended to supplement those in the Maintenance and minor repair tool kit. Included is a comprehensive set of sockets which, though expensive, are invaluable because of their versatility (especially when various extensions and drives are available). We recommend the 3/8 inch drive over the 1/2 inch drive for general vehicle maintenance and repair (ideally, the mechanic would have a 3/8 inch drive set and a 1/2 inch drive set).

Alternator rotor puller tool
Socket set(s)
Reversible ratchet
Extension - 6 in
Universal joint
Torque wrench (same size drive as sockets)
Ball peen hammer - 8 oz
Soft-faced hammer (plastic/rubber)
Standard screwdriver (1/4 in x 6 in)

Standard screwdriver (stubby - 5/16 in)
Phillips screwdriver (No. 3 x 8 in)
Phillips screwdriver (stubby - No. 2)
Pliers - locking
Pliers - lineman's
Pliers - needle nose
Pliers - snap-ring (internal and external)
Cold chisel - 1/2 in
Scriber
Scraper (made from flattened copper tubing)
Center punch
Pin punches (1/16, 1/8, 3/16 in)
Steel rule/straightedge - 12 in
Pin-type spanner wrench
A selection of files
Wire brush (large)

Note: *Another tool which is often useful is an electric drill with a chuck capacity of 3/8 inch (and a set of good quality drill bits).*

Special tools

The tools in this list include those which are not used regularly, are expensive to buy, or which need to be used in accordance with their manufacturer's instructions. Unless these tools will be used frequently, it is not very economical to purchase many of them. A consideration would be to split the cost and use between yourself and a friend or friends (i.e. members of a vehicle club).

This list primarily contains tools and instruments widely available to the public, as well as some special tools produced by the vehicle manufacturer for distribution to dealer service departments. As a result, references to the manufacturer's special tools are occasionally included in the text of this manual. Generally, an alternative method of doing the job without the special tool is offered. However, sometimes there is no alternative to their use. Where this is the case, and the tool can't be purchased or borrowed, the work should be turned over to the dealer service department or a vehicle repair shop.

Valve spring compressor
Piston ring removal and installation tool
Piston pin puller
Telescoping gauges
Micrometer(s) and/or dial/Vernier calipers
Cylinder surfacing hone
Cylinder compression gauge
Dial indicator set
Multimeter
Adjustable spanner
Manometer or vacuum gauge set
Small air compressor with blow gun and tire chuck

Buying tools

For the do-it-yourselfer who is just starting to get involved in vehicle maintenance and repair, there are a number of options available when purchasing tools. If maintenance and minor repair is the extent of the work to be done, the purchase of individual tools is satisfactory. If, on the other hand, extensive work is planned, it would be a good idea to purchase a modest tool set from one of the large retail chain stores. A set can usually be bought at a substantial savings over the individual tool prices (and they often come with a tool box). As additional tools are needed, add-on sets, individual tools and a larger tool box can be purchased to expand the tool selection. Building a tool set gradually allows the cost of the tools to be spread over a longer period of time and gives the mechanic the freedom to choose only those tools that will actually be used.

Tool stores and vehicle dealers will often be the only source of some of the special tools that are needed, but regardless of where tools are bought, try to avoid cheap ones (especially when buying screwdrivers and sockets) because they won't last very long. There are plenty of tools around at reasonable prices, but always aim to purchase items which meet the relevant national safety standards. The expense involved in replacing cheap tools will eventually be greater than the initial cost of quality tools.

It is obviously not possible to cover the subject of tools fully here. For those who wish to learn more about tools and their use, there is a book entitled *Vehicle Workshop Practice Manual* (Book no. 1454) available from the publishers of this manual. It also provides an introduction to basic workshop practice which will be of interest to a home mechanic working on any type of vehicle.

Care and maintenance of tools

Good tools are expensive, so it makes sense to treat them with respect. Keep them clean and in usable condition and store them properly when not in use. Always wipe off any dirt, grease or metal chips before putting them away. Never leave tools lying around in the work area.

Some tools, such as screwdrivers, pliers, wrenches and sockets, can be hung on a panel mounted on the garage or workshop wall, while others should be kept in a tool box or tray. Measuring instruments, gauges, meters, etc. must be carefully stored where they can't be damaged by weather or impact from other tools.

When tools are used with care and stored properly, they will last a very long time. Even with the best of care, tools will wear out if used frequently. When a tool is damaged or worn out, replace it; subsequent jobs will be safer and more enjoyable if you do.

Working facilities

Not to be overlooked when discussing tools is the workshop. If anything more than routine maintenance is to be carried out, some sort of suitable work area is essential.

It is understood, and appreciated, that many home mechanics do not have a good workshop or garage available and end up removing an engine or doing major repairs outside (it is recommended, however, that the overhaul or repair be completed under the cover of a roof).

A clean, flat workbench or table of comfortable working height is an absolute necessity. The workbench should be equipped with a vise that has a jaw opening of at least four inches.

As mentioned previously, some clean, dry storage space is also required for tools, as well as the lubricants, fluids, cleaning solvents, etc. which soon become necessary.

Sometimes waste oil and fluids, drained from the engine or cooling system during normal maintenance or repairs, present a disposal problem. To avoid pouring them on the ground or into a sewage system, simply pour the used fluids into large containers, seal them with caps and take them to an authorized disposal site or service station. Plastic jugs are ideal for this purpose.

Always keep a supply of old newspapers and clean rags available. Old towels are excellent for mopping up spills. Many mechanics use rolls of paper towels for most work because they are readily available and disposable. To help keep the area under the vehicle clean, a large cardboard box can be cut open and flattened to protect the garage or shop floor.

Whenever working over a painted surface (such as the fuel tank) cover it with an old blanket or bedspread to protect the finish.

Safety first

Professional mechanics are trained in safe working procedures. However enthusiastic you may be about getting on with the job at hand, take the time to ensure that your safety is not put at risk. A moment's lack of attention can result in an accident, as can failure to observe simple precautions.

There will always be new ways of having accidents, and the following is not a comprehensive list of all dangers; it is intended rather to make you aware of the risks and to encourage a safe approach to all work you carry out on your bike.

Essential DOs and DON'Ts

DON'T start the engine without first ascertaining that the transmission is in neutral.

DON'T attempt to drain oil until you are sure it has cooled sufficiently to avoid scalding you.

DON'T grasp any part of the engine or exhaust system without first ascertaining that it is cool enough not to burn you.

DON'T allow brake fluid to contact the machine's paint work or plastic components.

DON'T siphon toxic liquids such as fuel, hydraulic fluid or antifreeze by mouth, or allow them to remain on your skin.

DON'T inhale dust - it may be injurious to health (see *Asbestos* heading).

DON'T allow any spilled oil or grease to remain on the floor - wipe it up right away, before someone slips on it.

DON'T use ill fitting wrenches or other tools which may slip and cause injury.

DON'T attempt to lift a heavy component which may be beyond your capability - get assistance.

DON'T rush to finish a job or take unverified short cuts.

DON'T allow children or animals in or around an unattended vehicle.

DON'T inflate a tire to a pressure above the recommended maximum. Apart from over stressing the carcase and wheel rim, in extreme cases the tire may blow off forcibly. ATV tires, which are designed to operate at very low air pressures, may rupture if overinflated.

DO ensure that the machine is supported securely at all times. This is especially important when the machine is blocked up to aid wheel or suspension removal.

DO take care when attempting to loosen a stubborn nut or bolt. It is generally better to pull on a wrench, rather than push, so that if you slip, you fall away from the machine rather than onto it.

DO wear eye protection when using power tools such as drill, sander, bench grinder etc.

DO use a barrier cream on your hands prior to undertaking dirty jobs - it will protect your skin from infection as well as making the dirt easier to remove afterwards; but make sure your hands aren't left slippery. Note that long-term contact with used engine oil can be a health hazard.

DO keep loose clothing (cuffs, ties etc. and long hair) well out of the way of moving mechanical parts.

DO remove rings, wristwatch etc., before working on the vehicle- especially the electrical system.

DO keep your work area tidy - it is only too easy to fall over articles left lying around.

DO exercise caution when compressing springs for removal or installation. Ensure that the tension is applied and released in a controlled manner, using suitable tools which preclude the possibility of the spring escaping violently.

DO ensure that any lifting tackle used has a safe working load rating adequate for the job.

DO get someone to check periodically that all is well, when working alone on the vehicle.

DO carry out work in a logical sequence and check that everything is correctly assembled and tightened afterwards.

DO remember that your vehicle's safety affects that of yourself and others. If in doubt on any point, get professional advice.

IF, in spite of following these precautions, you are unfortunate enough to injure yourself, seek medical attention as soon as possible.

Asbestos

Certain friction, insulating, sealing and other products - such as brake pads, clutch linings, gaskets, etc. - may contain asbestos. *Extreme care must be taken to avoid inhalation of dust from such products since it is hazardous to health*. If in doubt, assume that they *do* contain asbestos.

Fire

Remember at all times that gasoline (petrol) is highly flammable. Never smoke or have any kind of naked flame around, when working on the vehicle. But the risk does not end there - a spark caused by an electrical short-circuit, by two metal surfaces contacting each other, by careless use of tools, or even by static electricity built up in your body under certain conditions, can ignite gasoline (petrol) vapor, which in a confined space is highly explosive. Never use gasoline (petrol) as a cleaning solvent. Use an approved safety solvent.

Always disconnect the battery ground (earth) terminal before working on any part of the fuel or electrical system, and never risk spilling fuel on to a hot engine or exhaust.

It is recommended that a fire extinguisher of a type suitable for fuel and electrical fires is kept handy in the garage or workplace at all times. Never try to extinguish a fuel or electrical fire with water.

Fumes

Certain fumes are highly toxic and can quickly cause unconsciousness and even death if inhaled to any extent. Gasoline (petrol) vapor comes into this category, as do the vapors from certain solvents such as trichloroethylene. Any draining or pouring of such volatile fluids should be done in a well ventilated area.

When using cleaning fluids and solvents, read the instructions carefully. Never use materials from unmarked containers - they may give off poisonous vapors.

Never run the engine of a motor vehicle in an enclosed space such as a garage. Exhaust fumes contain carbon monoxide which is extremely poisonous; if you need to run the engine, always do so in the open air or at least have the rear of the vehicle outside the workplace.

The battery

Never cause a spark, or allow a bare light bulb near the vehicle's battery. It will normally be giving off a certain amount of hydrogen gas, which is highly explosive.

Always disconnect the battery ground (earth) terminal before working on the fuel or electrical systems (except where noted).

Do not charge the battery at an excessive rate or the battery may burst.

Take care when cleaning or carrying the battery. The acid electrolyte, even when diluted, is very corrosive and should not be allowed to contact the eyes or skin. Always wear rubber gloves and goggles or a face shield. If you ever need to prepare electrolyte yourself, always add the acid slowly to the water; never add the water to the acid.

Electricity

When using an electric power tool, inspection light etc., always ensure that the appliance is correctly connected to its plug and that, where necessary, it is properly grounded (earthed). Do not use such appliances in damp conditions and, again, beware of creating a spark or applying excessive heat in the vicinity of fuel or fuel vapor. Also ensure that the appliances meet national safety standards.

A severe electric shock can result from touching certain parts of the electrical system, such as the spark plug wires (HT leads), when the engine is running or being cranked, particularly if components are damp or the insulation is defective. Where an electronic ignition system is used, the secondary (HT) voltage is much higher and could prove fatal.

ATV chemicals and lubricants

A number of chemicals and lubricants are available for use in vehicle maintenance and repair. They include a wide variety of products ranging from cleaning solvents and degreasers to lubricants and protective sprays for rubber, plastic and vinyl.

Contact point/spark plug cleaner is a solvent used to clean oily film and dirt from points, grime from electrical connectors and oil deposits from spark plugs. It is oil free and leaves no residue. It can also be used to remove gum and varnish from carburetor jets and other orifices.

Carburetor cleaner is similar to contact point/spark plug cleaner but it usually has a stronger solvent and may leave a slight oily residue. It is not recommended for cleaning electrical components or connections.

Brake system cleaner is used to remove grease or brake fluid from brake system components (where clean surfaces are absolutely necessary and petroleum-based solvents cannot be used); it also leaves no residue.

Silicone-based lubricants are used to protect rubber parts such as hoses and grommets, and are used as lubricants for hinges and locks.

Multi-purpose grease is an all purpose lubricant used wherever grease is more practical than a liquid lubricant such as oil. Some multi-purpose grease is colored white and specially formulated to be more resistant to water than ordinary grease.

Gear oil (sometimes called gear lube) is a specially designed oil used in transmissions and final drive units, as well as other areas where high friction, high temperature lubrication is required. It is available in a number of viscosities (weights) for various applications.

Motor oil, of course, is the lubricant specially formulated for use in the engine. It normally contains a wide variety of additives to prevent corrosion and reduce foaming and wear. Motor oil comes in various weights (viscosity ratings) of from 5 to 80. The recommended weight of the oil depends on the seasonal temperature and the demands on the engine. Light oil is used in cold climates and under light load conditions; heavy oil is used in hot climates and where high loads are encountered. Multi-viscosity oils are designed to have characteristics of both light and heavy oils and are available in a number of weights from 5W-20 to 20W-50. On these machines, the same oil supply is shared by the engine and transmission.

Gas (petrol) additives perform several functions, depending on their chemical makeup. They usually contain solvents that help dissolve gum and varnish that build up on carburetor and intake parts. They also serve to break down carbon deposits that form on the inside surfaces of the combustion chambers. Some additives contain upper cylinder lubricants for valves and piston rings.

Brake fluid is a specially formulated hydraulic fluid that can withstand the heat and pressure encountered in brake systems. Care must be taken that this fluid does not come in contact with painted surfaces or plastics. An opened container should always be resealed to prevent contamination by water or dirt.

Chain lubricants are formulated especially for use on the final drive chains of vehicles so equipped (all models covered in this manual are equipped with shaft drive). A good chain lube should adhere well and have good penetrating qualities to be effective as a lubricant inside the chain and on the side plates, pins and rollers. Most chain lubes are either the foaming type or quick drying type and are usually marketed as sprays.

Degreasers are heavy duty solvents used to remove grease and grime that may accumulate on engine and frame components. They can be sprayed or brushed on and, depending on the type, are rinsed with either water or solvent.

Solvents are used alone or in combination with degreasers to clean parts and assemblies during repair and overhaul. The home mechanic should use only solvents that are non-flammable and that do not produce irritating fumes.

Gasket sealing compounds may be used in conjunction with gaskets, to improve their sealing capabilities, or alone, to seal metal-to-metal joints. Many gasket sealers can withstand extreme heat, some are impervious to gasoline and lubricants, while others are capable of filling and sealing large cavities. Depending on the intended use, gasket sealers either dry hard or stay relatively soft and pliable. They are usually applied by hand, with a brush, or are sprayed on the gasket sealing surfaces.

Thread cement is an adhesive locking compound that prevents threaded fasteners from loosening because of vibration. It is available in a variety of types for different applications.

Moisture dispersants are usually sprays that can be used to dry out electrical components such as the fuse block and wiring connectors. Some types can also be used as treatment for rubber and as a lubricant for hinges, cables and locks.

Waxes and polishes are used to help protect painted and plated surfaces from the weather. Different types of paint may require the use of different types of wax polish. Some polishes utilize a chemical or abrasive cleaner to help remove the top layer of oxidized (dull) paint on older vehicles. In recent years, many non-wax polishes (that contain a wide variety of chemicals such as polymers and silicones) have been introduced. These non-wax polishes are usually easier to apply and last longer than conventional waxes and polishes.

Troubleshooting

Contents

Symptom   *Section*

Engine doesn't start or is difficult to start

Starter motor doesn't rotate.. 1
Starter motor rotates but engine does not turn over 2
Starter works but engine won't turn over (seized) 3
No fuel flow 4
Engine flooded 5
No spark or weak spark 6
Compression low 7
Stalls after starting 8
Rough idle 9

Poor running at low speed

Spark weak 10
Fuel/air mixture incorrect 11
Compression low 12
Poor acceleration 13

Poor running or no power at high speed

Firing incorrect 14
Fuel/air mixture incorrect 15
Compression low 16
Knocking or pinging 17
Miscellaneous causes 18

Overheating

Engine overheats... 19
Firing incorrect 20
Fuel/air mixture incorrect 21
Compression too high.. 22
Engine load excessive... 23
Lubrication inadequate 24
Miscellaneous causes 25

Clutch problems

Clutch slipping 26
Clutch not disengaging completely....................................... 27

Gear shifting problems

Doesn't go into gear, or lever doesn't return 28
Jumps out of gear 29
Overshifts 30

Symptom   *Section*

Abnormal engine noise

Knocking or pinging 31
Piston slap or rattling 32
Valve noise 33
Other noise 34

Abnormal driveline noise

Clutch noise 35
Transmission noise 36
Final drive noise 37

Abnormal frame and suspension noise

Front end noise 38
Shock absorber noise 39
Disc brake noise.. 40

Oil level indicator light comes on

Engine lubrication system 41
Electrical system 42

Excessive exhaust smoke

White smoke 43
Black smoke.. 44
Brown smoke 45

Poor handling or stability

Handlebar hard to turn... 46
Handlebar shakes or vibrates excessively 47
Handlebar pulls to one side 48
Poor shock absorbing qualities.. 49

Braking problems

Brakes are spongy, don't hold... 50
Brake lever pulsates 51
Brakes drag 52

Electrical problems

Battery dead or weak.. 53
Battery overcharged... 54

Engine doesn't start or is difficult to start

1   Starter motor does not rotate

1   Engine kill switch Off.
2   Fuse blown. Check fuse (Chapter 8).
3   Battery voltage low. Check and recharge battery (Chapter 8).
4   Starter motor defective. Make sure the wiring to the starter is secure. Test starter relay (Chapter 8). If the relay is good, then the fault is in the wiring or motor.
5   Starter relay faulty. Check it according to the procedure in Chapter 8.
6   Starter switch not contacting. The contacts could be wet, corroded or dirty. Disassemble and clean the switch (Chapter 8).
7   Wiring open or shorted. Check all wiring connections and harnesses to make sure that they are dry, tight and not corroded. Also check for broken or frayed wires that can cause a short to ground (see wiring diagram, Chapter 8).
8   Ignition (main) switch defective. Check the switch according to the procedure in Chapter 8. Replace the switch with a new one if it is defective.
9   Engine kill switch defective. Check for wet, dirty or corroded contacts. Clean or replace the switch as necessary (Chapter 8).
10   Starter diode defective. Test as described in Chapter 8.

2   Starter motor rotates but engine does not turn over

1   Starter motor clutch defective. Inspect and repair or replace (Chapter 8).
2   Damaged starter reduction gears. Inspect and replace the damaged parts (Chapter 2).

3   Starter works but engine won't turn over (seized)

Seized engine caused by one or more internally damaged components. Failure due to wear, abuse or lack of lubrication. Damage can include seized valves, valve lifters, camshaft, piston, crankshaft, connecting rod bearings, or transmission gears or bearings. Refer to Chapter 2 for engine disassembly.

4   No fuel flow

1   No fuel in tank.
2   Tank cap air vent obstructed. Usually caused by dirt or water. Remove it and clean the cap vent hole.
3   Clogged strainer in fuel tap or inside fuel tank. Remove and clean the strainer(s) (Chapter 1).
4   Fuel line clogged. Pull the fuel line loose and carefully blow through it.
5   Inlet needle valve clogged. A very bad batch of fuel with an unusual additive may have been used, or some other foreign material has entered the tank. Many times after a machine has been stored for many months without running, the fuel turns to a varnish-like liquid and forms deposits on the inlet needle valve and jets. The carburetor should be removed and overhauled if draining the float chamber doesn't solve the problem.

5   Engine flooded

1   Float level too high. Check as described in Chapter 3 and replace the float if necessary.
2   Inlet needle valve worn or stuck open. A piece of dirt, rust or other debris can cause the inlet needle to seat improperly, causing excess fuel to be admitted to the float bowl. In this case, the float chamber should be cleaned and the needle and seat inspected. If the needle and seat are worn, then the leaking will persist and the parts should be replaced with new ones (Chapter 3).
3   Starting technique incorrect. Under normal circumstances (i.e., if all the carburetor functions are sound) the machine should start with little or no throttle. When the engine is cold, the choke should be operated and the engine started without opening the throttle. When the engine is at operating temperature, only a very slight amount of throttle should be necessary. If the engine is flooded, turn the fuel tap off and hold the throttle open while cranking the engine. This will allow additional air to reach the cylinder. Remember to turn the fuel tap back on after the engine starts.

6   No spark or weak spark

1   Ignition switch Off.
2   Engine kill switch turned to the Off position.
3   Battery voltage low. Check and recharge battery as necessary (Chapter 8).
4   Spark plug dirty, defective or worn out. Locate reason for fouled plug using spark plug condition chart and follow the plug maintenance procedures in Chapter 1.
5   Spark plug cap or secondary (HT) wiring faulty. Check condition. Replace either or both components if cracks or deterioration are evident (Chapter 4).
6   Spark plug cap not making good contact. Make sure that the plug cap fits snugly over the plug end.
7   Ignition control module defective. Check the unit, referring to Chapter 4 for details.
8   Ignition signal generator defective. Check the unit, referring to Chapter 4 for details.
9   Ignition coil defective. Check the coil, referring to Chapter 4.
10   Ignition or kill switch shorted. This is usually caused by water, corrosion, damage or excessive wear. The kill switch can be disassembled and cleaned with electrical contact cleaner. If cleaning does not help, replace the switches (Chapter 8).
11   Wiring shorted or broken between:
   a) *Ignition switch and engine kill switch (or blown fuse)*
   b) *Ignition control unit and engine kill switch*
   c) *Ignition control unit and ignition coil*
   d) *Ignition coil and plug*
   e) *Ignition control unit and ignition signal generator*

Make sure that all wiring connections are clean, dry and tight. Look for chafed and broken wires (Chapters 4 and 8).

7   Compression low

1   Spark plug loose. Remove the plug and inspect the threads. Reinstall and tighten to the specified torque (Chapter 1).
2   Cylinder head not sufficiently tightened down. If the cylinder head is suspected of being loose, then there's a chance that the gasket or head is damaged if the problem has persisted for any length of time. The head nuts and bolts should be tightened to the proper torque in the correct sequence (Chapter 2).
3   Improper valve clearance. This means that the valve is not closing completely and compression pressure is leaking past the valve. Check and adjust the valve clearances (Chapter 1).
4   Cylinder and/or piston worn. Excessive wear will cause compression pressure to leak past the rings. This is usually accompanied by worn rings as well. A top end overhaul is necessary (Chapter 2).
5   Piston rings worn, weak, broken, or sticking. Broken or sticking piston rings usually indicate a lubrication or carburetion problem that causes excess carbon deposits or seizures to form on the pistons and rings. Top end overhaul is necessary (Chapter 2).
6   Piston ring-to-groove clearance excessive. This is caused by

excessive wear of the piston ring lands. Piston replacement is necessary (Chapter 2).

7   Cylinder head gasket damaged. If the head is allowed to become loose, or if excessive carbon build-up on a piston crown and combustion chamber causes extremely high compression, the head gasket may leak. Retorquing the head is not always sufficient to restore the seal, so gasket replacement is necessary (Chapter 2).

8   Cylinder head warped. This is caused by overheating or improperly tightened head nuts and bolts. Machine shop resurfacing or head replacement is necessary (Chapter 2).

9   Valve spring broken or weak. Caused by component failure or wear; the spring(s) must be replaced (Chapter 2).

10   Valve not seating properly. This is caused by a bent valve (from over-revving or improper valve adjustment), burned valve or seat (improper carburetion) or an accumulation of carbon deposits on the seat (from carburetion or lubrication problems). The valves must be cleaned and/or replaced and the seats serviced if possible (Chapter 2).

8   Stalls after starting

1   Improper choke action. Make sure the choke cable is getting a full stroke and staying in the out position.
2   Ignition malfunction. See Chapter 4.
3   Carburetor malfunction. See Chapter 3.
4   Fuel contaminated. The fuel can be contaminated with either dirt or water, or can change chemically if the machine is allowed to sit for several months or more. Drain the tank and float bowl and refill with fresh fuel (Chapter 3).
5   Intake air leak. Check for loose carburetor-to-intake joint connections or loose carburetor top (Chapter 3).
6   Engine idle speed incorrect. Turn throttle stop screw until the engine idles at the specified rpm (Chapter 1).

9   Rough idle

1   Ignition malfunction. See Chapter 4.
2   Idle speed incorrect. See Chapter 1.
3   Carburetor malfunction. See Chapter 3.
4   Idle fuel/air mixture incorrect. See Chapter 3.
5   Fuel contaminated. The fuel can be contaminated with either dirt or water, or can change chemically if the machine is allowed to sit for several months or more. Drain the tank and float bowls (Chapter 3).
6   Intake air leak. Check for loose carburetor-to-intake joint connections, loose or missing vacuum gauge access port cap or hose, or loose carburetor top (Chapter 3).
7   Air cleaner clogged. Service or replace air cleaner element (Chapter 1).

Poor running at low speed

10   Spark weak

1   Battery voltage low. Check and recharge battery (Chapter 8).
2   Spark plug fouled, defective or worn out. Refer to Chapter 1 for spark plug maintenance.
3   Spark plug cap or secondary (HT) wiring defective. Refer to Chapters 1 and 4 for details on the ignition system.
4   Spark plug cap not making contact.
5   Incorrect spark plug. Wrong type, heat range or cap configuration. Check and install correct plug listed in Chapter 1. A cold plug or one with a recessed firing electrode will not operate at low speeds without fouling.
6   Ignition control module defective. See Chapter 4.
7   Signal generator defective. See Chapter 4.
8   Ignition coil defective. See Chapter 4.

11   Fuel/air mixture incorrect

1   Pilot screw out of adjustment (Chapter 3).
2   Pilot jet or air passage clogged. Remove and overhaul the carburetor (Chapter 3).
3   Air bleed holes clogged. Remove carburetor and blow out all passages (Chapter 3).
4   Air cleaner clogged, poorly sealed or missing.
5   Air cleaner-to-carburetor boot poorly sealed. Look for cracks, holes or loose clamps and replace or repair defective parts.
6   Float level too high or too low. Check and replace the float if necessary (Chapter 3).
7   Fuel tank air vent obstructed. Make sure that the air vent passage in the filler cap is open.
8   Carburetor intake joint loose. Check for cracks, breaks, tears or loose clamps or bolts. Repair or replace the rubber boot and its O-ring.

12   Compression low

1   Spark plug loose. Remove the plug and inspect the threads. Reinstall and tighten to the specified torque (Chapter 1).
2   Cylinder head not sufficiently tightened down. If the cylinder head is suspected of being loose, then there's a chance that the gasket and head are damaged if the problem has persisted for any length of time. The head nuts and bolts should be tightened to the proper torque in the correct sequence (Chapter 2).
3   Improper valve clearance. This means that the valve is not closing completely and compression pressure is leaking past the valve. Check and adjust the valve clearances (Chapter 1).
4   Cylinder and/or piston worn. Excessive wear will cause compression pressure to leak past the rings. This is usually accompanied by worn rings as well. A top end overhaul is necessary (Chapter 2).
5   Piston rings worn, weak, broken, or sticking. Broken or sticking piston rings usually indicate a lubrication or carburetion problem that causes excess carbon deposits or seizures to form on the pistons and rings. Top end overhaul is necessary (Chapter 2).
6   Piston ring-to-groove clearance excessive. This is caused by excessive wear of the piston ring lands. Piston replacement is necessary (Chapter 2).
7   Cylinder head gasket damaged. If the head is allowed to become loose, or if excessive carbon build-up on the piston crown and combustion chamber causes extremely high compression, the head gasket may leak. Retorquing the head is not always sufficient to restore the seal, so gasket replacement is necessary (Chapter 2).
8   Cylinder head warped. This is caused by overheating or improperly tightened head nuts and bolts. Machine shop resurfacing or head replacement is necessary (Chapter 2).
9   Valve spring broken or weak. Caused by component failure or wear; the spring(s) must be replaced (Chapter 2).
10   Valve not seating properly. This is caused by a bent valve (from over-revving or improper valve adjustment), burned valve or seat (improper carburetion) or an accumulation of carbon deposits on the seat (from carburetion, lubrication problems). The valves must be cleaned and/or replaced and the seats serviced if possible (Chapter 2).

13   Poor acceleration

1   Carburetor leaking or dirty. Overhaul the carburetor (Chapter 3).
2   Timing not advancing. The signal generator or the ignition control module may be defective. If so, they must be replaced with new ones, as they can't be repaired.
3   Engine oil viscosity too high. Using a heavier oil than that recommended in Chapter 1 can damage the oil pump or lubrication system and cause drag on the engine.
4   Brakes dragging. Usually caused by debris which has entered the

brake piston sealing boots (front brakes), corroded wheel cylinders (front brakes), sticking brake cam (rear brakes) or from a warped drum or bent axle. Repair as necessary (Chapter 6).

Poor running or no power at high speed

14 Firing incorrect

1   Air cleaner restricted. Clean or replace element (Chapter 1).
2   Spark plug fouled, defective or worn out. See Chapter 1 for spark plug maintenance.
3   Spark plug cap or secondary (HT) wiring defective. See Chapters 1 and 4 for details of the ignition system.
4   Spark plug cap not in good contact. See Chapter 4.
5   Incorrect spark plug. Wrong type, heat range or cap configuration. Check and install correct plugs listed in Chapter 1. A cold plug or one with a recessed firing electrode will not operate at low speeds without fouling.
6   Ignition control module defective. See Chapter 4.
7   Ignition coil defective. See Chapter 4.

15 Fuel/air mixture incorrect

1   Pilot screw out of adjustment. See Chapter 3 for adjustment procedures.
2   Main jet clogged. Dirt, water or other contaminants can clog the main jets. Clean the fuel tap strainer and in-tank strainer, the float bowl area, and the jets and carburetor orifices (Chapter 3).
3   Main jet wrong size. The standard jetting is for sea level atmospheric pressure and oxygen content. See Chapter 3 for high altitude adjustments.
4   Throttle shaft-to-carburetor body clearance excessive. Refer to Chapter 3 for inspection and part replacement procedures.
5   Air bleed holes clogged. Remove and overhaul carburetor (Chapter 3).
6   Air cleaner clogged, poorly sealed, or missing.
7   Air cleaner-to-carburetor boot poorly sealed. Look for cracks, holes or loose clamps, and replace or repair defective parts.
8   Float level too high or too low. Check float level and replace the float if necessary (Chapter 3).
9   Fuel tank air vent obstructed. Make sure the air vent passage in the filler cap is open.
10   Carburetor intake joint loose. Check for cracks, breaks, tears or loose clamps or bolts. Repair or replace the rubber boots (Chapter 3).
11   Fuel tap clogged. Remove the tap and clean it (Chapter 1).
12   Fuel line clogged. Pull the fuel line loose and carefully blow through it.

16 Compression low

1   Spark plug loose. Remove the plug and inspect the threads. Reinstall and tighten to the specified torque (Chapter 1).
2   Cylinder head not sufficiently tightened down. If the cylinder head is suspected of being loose, then there's a chance that the gasket and head are damaged if the problem has persisted for any length of time. The head nuts and bolts should be tightened to the proper torque in the correct sequence (Chapter 2).
3   Improper valve clearance. This means that the valve is not closing completely and compression pressure is leaking past the valve. Check and adjust the valve clearances (Chapter 1).
4   Cylinder and/or piston worn. Excessive wear will cause compression pressure to leak past the rings. This is usually accompanied by worn rings as well. A top end overhaul is necessary (Chapter 2).
5   Piston rings worn, weak, broken, or sticking. Broken or sticking piston rings usually indicate a lubrication or carburetion problem that

causes excess carbon deposits or seizures to form on the pistons and rings. Top end overhaul is necessary (Chapter 2).
6   Piston ring-to-groove clearance excessive. This is caused by excessive wear of the piston ring lands. Piston replacement is necessary (Chapter 2).
7   Cylinder head gasket damaged. If a head is allowed to become loose, or if excessive carbon build-up on the piston crown and combustion chamber causes extremely high compression, the head gasket may leak. Retorquing the head is not always sufficient to restore the seal, so gasket replacement is necessary (Chapter 2).
8   Cylinder head warped. This is caused by overheating or improperly tightened head nuts and bolts. Machine shop resurfacing or head replacement is necessary (Chapter 2).
9   Valve spring broken or weak. Caused by component failure or wear; the spring(s) must be replaced (Chapter 2).
10   Valve not seating properly. This is caused by a bent valve (from over-revving or improper valve adjustment), burned valve or seat (improper carburetion) or an accumulation of carbon deposits on the seat (from carburetion or lubrication problems). The valves must be cleaned and/or replaced and the seats serviced if possible (Chapter 2).

17 Knocking or pinging

1   Carbon build-up in combustion chamber. Use of a fuel additive that will dissolve the adhesive bonding the carbon particles to the crown and chamber is the easiest way to remove the build-up. Otherwise, the cylinder head will have to be removed and decarbonized (Chapter 2).
2   Incorrect or poor quality fuel. Old or improper grades of fuel can cause detonation. This causes the piston to rattle, thus the knocking or pinging sound. Drain old fuel and always use the recommended fuel grade.
3   Spark plug heat range incorrect. Uncontrolled detonation indicates the plug heat range is too hot. The plug in effect becomes a glow plug, raising cylinder temperatures. Install the proper heat range plug (Chapter 1).
4   Improper air/fuel mixture. This will cause the cylinder to run hot, which leads to detonation. Clogged jets or an air leak can cause this imbalance. See Chapter 3.

18 Miscellaneous causes

1   Throttle valve doesn't open fully. Adjust the cable slack (Chapter 1).
2   Clutch slipping. May be caused by improperly adjustment or loose or worn clutch components. Refer to Chapter 1 for adjustment or Chapter 2 for cable replacement and clutch overhaul procedures.
3   Timing not advancing.
4   Engine oil viscosity too high. Using a heavier oil than the one recommended in Chapter 1 can damage the oil pump or lubrication system and cause drag on the engine.
5   Brakes dragging. Usually caused by debris which has entered the brake piston sealing boot, or from a warped disc or bent axle. Repair as necessary.

Overheating

19 Engine overheats

1   Engine oil level low. Check and add oil (Chapter 1).
2   Wrong type of oil. If you're not sure what type of oil is in the engine, drain it and fill with the correct type (Chapter 1).
3   Air leak at carburetor intake joints. Check and tighten or replace as necessary (Chapter 3).
4   Fuel level low. Check and adjust if necessary (Chapter 3).

5   Worn oil pump or clogged oil passages. Replace pump or clean passages as necessary.
6   Clogged external oil line. Remove and check for foreign material (see Chapter 2).
7   Carbon build-up in combustion chambers. Use of a fuel additive that will dissolve the adhesive bonding the carbon particles to the piston crown and chambers is the easiest way to remove the build-up. Otherwise, the cylinder head will have to be removed and decarbonized (Chapter 2).
8   Operation in high ambient temperatures. A cooling fan kit, available from Honda dealers, can be installed.

20   Firing incorrect

1   Spark plug fouled, defective or worn out. See Chapter 1 for spark plug maintenance.
2   Incorrect spark plug (see Chapter 1).
3   Faulty ignition coil(s) (Chapter 4).

21   Fuel/air mixture incorrect

1   Pilot screw out of adjustment (Chapter 3).
2   Main jet clogged. Dirt, water and other contaminants can clog the main jets. Clean the fuel tap strainer and in-tank strainer, the float bowl area and the jets and carburetor orifices (Chapter 3).
3   Main jet wrong size. The standard jetting is for sea level atmospheric pressure and oxygen content. See Chapter 3 for high altitude settings.
4   Air cleaner poorly sealed or missing.
5   Air cleaner-to-carburetor boot poorly sealed. Look for cracks, holes or loose clamps and replace or repair.
6   Fuel level too low. Check float level and replace the float if necessary (Chapter 3).
7   Fuel tank air vent obstructed. Make sure that the air vent passage in the filler cap is open.
8   Carburetor intake joint loose. Check for cracks, breaks, tears or loose clamps or bolts. Repair or replace the rubber boot and its O-ring (Chapter 3).

22   Compression too high

1   Carbon build-up in combustion chamber. Use of a fuel additive that will dissolve the adhesive bonding the carbon particles to the piston crown and chamber is the easiest way to remove the build-up. Otherwise, the cylinder head will have to be removed and decarbonized (Chapter 2).
2   Improperly machined head surface or installation of incorrect gasket during engine assembly.

23   Engine load excessive

1   Clutch slipping. Can be caused by damaged, loose or worn clutch components. Refer to Chapter 2 for overhaul procedures.
2   Engine oil level too high. The addition of too much oil will cause pressurization of the crankcase and inefficient engine operation. Check Specifications and drain to proper level (Chapter 1).
3   Engine oil viscosity too high. Using a heavier oil than the one recommended in Chapter 1 can damage the oil pump or lubrication system as well as cause drag on the engine.
4   Brakes dragging. Usually caused by debris which has entered the brake piston sealing boots (front brakes), corroded wheel cylinders (front brakes), sticking brake cam (rear brakes) or from a warped drum or bent axle. Repair as necessary (Chapter 6).

24   Lubrication inadequate

1   Engine oil level too low. Friction caused by intermittent lack of lubrication or from oil that is overworked can cause overheating. The oil provides a definite cooling function in the engine. Check the oil level (Chapter 1).
2   Poor quality engine oil or incorrect viscosity or type. Oil is rated not only according to viscosity but also according to type. Some oils are not rated high enough for use in this engine. Check the Specifications section and change to the correct oil (Chapter 1).
3   Camshaft or journals worn. Excessive wear causing drop in oil pressure. Replace cam or cylinder head. Abnormal wear could be caused by oil starvation at high rpm from low oil level or improper viscosity or type of oil (Chapter 1).
4   Crankshaft and/or bearings worn. Same problems as paragraph 3. Check and replace crankshaft assembly if necessary (Chapter 2).

25   Miscellaneous causes

Modification to exhaust system. Most aftermarket exhaust systems cause the engine to run leaner, which makes it run hotter. When installing an accessory exhaust system, always rejet the carburetor.

Clutch problems

26   Clutch slipping

1   Change clutch friction plates worn or warped. Overhaul the change clutch assembly (Chapter 2).
2   Change clutch steel plates worn or warped (Chapter 2).
3   Change clutch spring(s) broken or weak. Old or heat-damaged spring(s) (from slipping clutch) should be replaced with new ones (Chapter 2).
4   Change clutch release mechanism defective. Replace any defective parts (Chapter 2).
5   Change clutch center or housing unevenly worn. This causes improper engagement of the plates. Replace the damaged or worn parts (Chapter 2).
6   Centrifugal clutch weight linings or drum worn (Chapter 2).

27   Clutch not disengaging completely

1   Change clutch improperly adjusted (see Chapter 1).
2   Change clutch plates warped or damaged. This will cause clutch drag, which in turn will cause the machine to creep. Overhaul the clutch assembly (Chapter 2).
3   Sagged or broken change clutch spring(s). Check and replace the spring(s) (Chapter 2).
4   Engine oil deteriorated. Old, thin, worn out oil will not provide proper lubrication for the discs, causing the change clutch to drag. Replace the oil and filter (Chapter 1).
5   Engine oil viscosity too high. Using a thicker oil than recommended in Chapter 1 can cause the change clutch plates to stick together, putting a drag on the engine. Change to the correct viscosity oil (Chapter 1).
6   Change clutch housing seized on shaft. Lack of lubrication, severe wear or damage can cause the housing to seize on the shaft. Overhaul of the clutch, and perhaps transmission, may be necessary to repair the damage (Chapter 2).
7   Change clutch release mechanism defective. Worn or damaged release mechanism parts can stick and fail to apply force to the pressure plate. Overhaul the release mechanism (Chapter 2).
8   Loose change clutch center nut. Causes housing and center

misalignment putting a drag on the engine. Engagement adjustment continually varies. Overhaul the clutch assembly (Chapter 2).

9   Weak or broken centrifugal clutch springs (Chapter 2).

Gear shifting problems

28   Doesn't go into gear or lever doesn't return

1   Clutch not disengaging. See Section 27.

2   Shift fork(s) bent or seized. May be caused by lack of lubrication. Overhaul the transmission (Chapter 2).

3   Gear(s) stuck on shaft. Most often caused by a lack of lubrication or excessive wear in transmission bearings and bushings. Overhaul the transmission (Chapter 2).

4   Shift drum binding. Caused by lubrication failure or excessive wear. Replace the drum and bearing (Chapter 2).

5   Shift lever return spring weak or broken (Chapter 2).

6   Shift lever broken. Splines stripped out of lever or shaft, caused by allowing the lever to get loose. Replace necessary parts (Chapter 2).

7   Shift mechanism pawl broken or worn. Full engagement and rotary movement of shift drum results. Replace shaft assembly (Chapter 2).

8   Pawl spring broken. Allows pawl to float, causing sporadic shift operation. Replace spring (Chapter 2).

29   Jumps out of gear

1   Shift fork(s) worn. Overhaul the transmission (Chapter 2).

2   Gear groove(s) worn. Overhaul the transmission (Chapter 2).

3   Gear dogs or dog slots worn or damaged. The gears should be inspected and replaced. No attempt should be made to service the worn parts.

30   Overshifts

1   Pawl spring weak or broken (Chapter 2).

2   Shift drum stopper lever not functioning (Chapter 2).

Abnormal engine noise

31   Knocking or pinging

1   Carbon build-up in combustion chamber. Use of a fuel additive that will dissolve the adhesive bonding the carbon particles to the piston crown and chamber is the easiest way to remove the build-up. Otherwise, the cylinder head will have to be removed and decarbonized (Chapter 2).

2   Incorrect or poor quality fuel. Old or improper fuel can cause detonation. This causes the pistons to rattle, thus the knocking or pinging sound. Drain the old fuel (Chapter 3) and always use the recommended grade fuel (Chapter 1).

3   Spark plug heat range incorrect. Uncontrolled detonation indicates that the plug heat range is too hot. The plug in effect becomes a glow plug, raising cylinder temperatures. Install the proper heat range plug (Chapter 1).

4   Improper air/fuel mixture. This will cause the cylinder to run hot and lead to detonation. Clogged jets or an air leak can cause this imbalance. See Chapter 3.

32   Piston slap or rattling

1   Cylinder-to-piston clearance excessive. Caused by improper

assembly. Inspect and overhaul top end parts (Chapter 2).

2   Connecting rod bent. Caused by over-revving, trying to start a badly flooded engine or from ingesting a foreign object into the combustion chamber. Replace the damaged parts (Chapter 2).

3   Piston pin or piston pin bore worn or seized from wear or lack of lubrication. Replace damaged parts (Chapter 2).

4   Piston ring(s) worn, broken or sticking. Overhaul the top end (Chapter 2).

5   Piston seizure damage. Usually from lack of lubrication or overheating. Replace the pistons and bore the cylinder, as necessary (Chapter 2).

6   Connecting rod upper or lower end clearance excessive. Caused by excessive wear or lack of lubrication. Replace worn parts.

33   Valve noise

1   Incorrect valve clearances. Adjust the clearances by referring to Chapter 1.

2   Valve spring broken or weak. Check and replace weak valve springs (Chapter 2).

3   Camshaft or cylinder head worn or damaged. Lack of lubrication at high rpm is usually the cause of damage. Insufficient oil or failure to change the oil at the recommended intervals are the chief causes.

34   Other noise

1   Cylinder head gasket leaking.

2   Exhaust pipe leaking at cylinder head connection. Caused by improper fit of pipe, damaged gasket or loose exhaust flange. All exhaust fasteners should be tightened evenly and carefully. Failure to do this will lead to a leak.

3   Crankshaft runout excessive. Caused by a bent crankshaft (from over-revving) or damage from an upper cylinder component failure.

4   Engine mounting bolts or nuts loose. Tighten all engine mounting bolts and nuts to the specified torque (Chapter 2).

5   Crankshaft bearings worn (Chapter 2).

6   Camshaft chain tensioner defective. Replace according to the procedure in Chapter 2.

7   Camshaft chain, sprockets or guides worn (Chapter 2).

Abnormal driveline noise

35   Clutch noise

1   Change clutch housing/friction plate clearance excessive (Chapter 2).

2   Loose or damaged change clutch pressure plate and/or bolts (Chapter 2).

3   Broken centrifugal clutch springs (Chapter 2).

36   Transmission noise

1   Bearings worn. Also includes the possibility that the shafts are worn. Overhaul the transmission (Chapter 2).

2   Gears worn or chipped (Chapter 2).

3   Metal chips jammed in gear teeth. Probably pieces from a broken clutch, gear or shift mechanism that were picked up by the gears. This will cause early bearing failure (Chapter 2).

4   Engine oil level too low. Causes a howl from transmission. Also affects engine power and clutch operation (Chapter 1).

37   Transfer case noise

1   Bearings worn. Also includes the possibility that the shafts are

worn. Overhaul the transfer case (Chapter 6).
2   Gears worn or chipped (Chapter 6).
3   Metal chips jammed in gear teeth. This will cause early bearing failure (Chapter 6).
4   Engine oil level too low. Causes a howl from transmission. Also affects engine power and clutch operation (Chapter 1).

38   Final drive noise

1   Final drive oil level low (Chapter 1).
2   Final drive gear lash out of adjustment. Checking and adjustment require special tools and skills and should be done by a Honda dealer.
3   Final drive gears damaged or worn. Overhaul requires special tools and skills and should be done by a Honda dealer.

Abnormal chassis noise

38   Suspension noise

1   Spring weak or broken. Makes a clicking or scraping sound.
2   Steering shaft bearings worn or damaged. Clicks when braking. Check and replace as necessary (Chapter 5).
3   Shock absorber fluid level incorrect. Indicates a leak caused by defective seal. Shock will be covered with oil. Replace shock (Chapter 5).
4   Defective shock absorber with internal damage. This is in the body of the shock and can't be remedied. The shock must be replaced with a new one (Chapter 5).
5   Bent or damaged shock body. Replace the shock with a new one (Chapter 5).

39   Driveaxle noise (4wd models)

1   Worn or damaged outer joint. Makes clicking noise in turns. Check for cut or damaged seals and repair as necessary (see Chapter 5).
2   Worn or damaged inner joint. Makes knock or clunk when accelerating after coasting. Check for cut or damaged seals and repair as necessary (see Chapter 5).

40   Brake noise

1   Brake linings worn or contaminated. Can cause scraping or squealing. Replace the shoes (Chapter 6).
2   Brake linings warped or worn unevenly. Can cause chattering. Replace the linings (Chapter 6).
3   Brake drum out of round. Can cause chattering. Replace brake drum (Chapter 6).
7   Loose or worn knuckle or rear axle bearings. Check and replace as needed (Chapter 5).

Oil temperature indicator light comes on

41   Engine lubrication system

1   High oil temperature due to operation in high ambient temperatures. Shut the engine off and let it cool. If the light comes on only due to operation in high ambient temperatures and not because of a mechanical problem, a cooling fan kit is available from Honda dealers.
2   Engine oil level low. Inspect for leak or other problem causing low oil level and add recommended oil (Chapters 1 and 2).

42   Electrical system

1   Oil temperature sensor defective. Check the sensor according to the procedure in Chapter 8. Replace it if it's defective.
2   Oil temperature indicator light circuit defective. Check for pinched, shorted, disconnected or damaged wiring (Chapter 8).
3   Oil temperature alarm unit defective. If no other cause can be found, the alarm unit may be at fault. Have it tested by a Honda dealer or substitute an alarm unit known to be good (Chapter 8).

Excessive exhaust smoke

43   White smoke

1   Piston oil ring worn. The ring may be broken or damaged, causing oil from the crankcase to be pulled past the piston into the combustion chamber. Replace the rings with new ones (Chapter 2).
2   Cylinders worn, cracked, or scored. Caused by overheating or oil starvation. If worn or scored, the cylinders will have to be rebored and new pistons installed. If cracked, the cylinder block will have to be replaced (see Chapter 2).
3   Valve oil seal damaged or worn. Replace oil seals with new ones (Chapter 2).
4   Valve guide worn. Perform a complete valve job (Chapter 2).
5   Engine oil level too high, which causes the oil to be forced past the rings. Drain oil to the proper level (Chapter 1).
6   Head gasket broken between oil return and cylinder. Causes oil to be pulled into the combustion chamber. Replace the head gasket and check the head for warpage (Chapter 2).
7   Abnormal crankcase pressurization, which forces oil past the rings. Clogged breather or hoses usually the cause (Chapter 2).

44   Black smoke

1   Air cleaner clogged. Clean or replace the element (Chapter 1).
2   Main jet too large or loose. Compare the jet size to the Specifications (Chapter 3).
3   Choke stuck, causing fuel to be pulled through choke circuit (Chapter 3).
4   Fuel level too high. Check the float level and replace the float if necessary (Chapter 3).
5   Inlet needle held off needle seat. Clean the float chamber and fuel line and replace the needle and seat if necessary (Chapter 3).

45   Brown smoke

1   Main jet too small or clogged. Lean condition caused by wrong size main jet or by a restricted orifice. Clean float chamber and jets and compare jet size to Specifications (Chapter 3).
2   Fuel flow insufficient. Fuel inlet needle valve stuck closed due to chemical reaction with old fuel. Float level incorrect; check and replace float if necessary. Restricted fuel line. Clean line and float chamber.
3   Carburetor intake tube loose (Chapter 3).
4   Air cleaner poorly sealed or not installed (Chapter 1).

Poor handling or stability

46   Handlebar hard to turn

1   Steering shaft nut too tight (Chapter 5).
2   Lower bearing or upper bushing damaged. Roughness can be felt as the bars are turned from side-to-side. Replace bearing and bushing (Chapter 5).

3   Steering shaft bearing lubrication inadequate. Caused by grease getting hard from age or being washed out by high pressure car washes. Remove steering shaft and replace bearing (Chapter 5).
4   Steering shaft bent. Caused by a collision, hitting a pothole or by rolling the machine. Replace damaged part. Don't try to straighten the steering shaft (Chapter 5).
5   Front tire air pressure too low (Chapter 1).

47   Handlebar shakes or vibrates excessively

1   Tires worn or out of balance (Chapter 1 or 6).
2   Swingarm bearings worn. Replace worn bearings by referring to Chapter 6.
3   Wheel rim(s) warped or damaged. Inspect wheels (Chapter 6).
4   Wheel bearings worn. Worn front or rear wheel bearings can cause poor tracking. Worn front bearings will cause wobble (Chapter 6).
5   Wheel hubs installed incorrectly (Chapter 5 or Chapter 6).
6   Handlebar clamp bolts or bracket nuts loose (Chapter 5).
7   Steering shaft nut or bolts loose. Tighten them to the specified torque (Chapter 5).
8   Motor mount bolts loose. Will cause excessive vibration with increased engine rpm (Chapter 2).

48   Handlebar pulls to one side

1   Uneven tire pressures (Chapter 1).
2   Frame bent. Definitely suspect this if the machine has been rolled. May or may not be accompanied by cracking near the bend. Replace the frame (Chapter 5).
3   Wheel out of alignment. Caused by incorrect toe-in adjustment (Chapter 1) or bent tie-rod (Chapter 5).
4   Swingarm bent or twisted. Caused by age (metal fatigue) or impact damage. Replace the swingarm (Chapter 5).
5   Steering shaft bent. Caused by impact damage or by rolling the vehicle. Replace the steering stem (Chapter 5).

49   Poor shock absorbing qualities

1   Too hard:
  a) *Shock internal damage.*
  b) *Tire pressure too high (Chapters 1 and 6).*
2 Too soft:
  a) *Shock oil insufficient and/or leaking (Chapter 5).*
  d) *Fork springs weak or broken (Chapter 5).*

Braking problems

50   Front brakes are spongy, don't hold

1   Air in brake line. Caused by inattention to master cylinder fluid level or by leakage. Locate problem and bleed brakes (Chapter 6).
2   Linings worn (Chapters 1 and 6).
3   Brake fluid leak. See paragraph 1.
4   Contaminated linings. Caused by contamination with oil, grease,
brake fluid, etc. Clean or replace linings. Clean drum thoroughly with brake cleaner (Chapter 6).
5   Brake fluid deteriorated. Fluid is old or contaminated. Drain system, replenish with new fluid and bleed the system (Chapter 6).
6   Master cylinder internal parts worn or damaged causing fluid to bypass (Chapter 6).
7   Master cylinder bore scratched by foreign material or broken spring. Repair or replace master cylinder (Chapter 6).
8   Drum warped. Replace drum (Chapter 6).

51   Brake lever or pedal pulsates

1   Axle bent. Replace axle (Chapter 5).
2   Wheel warped or otherwise damaged (Chapter 6).
3   Hub or axle bearings damaged or worn (Chapter 6).
4   Brake drum out of round. Replace brake drum (Chapter 6).

52   Brakes drag

1   Master cylinder piston seized. Caused by wear or damage to piston or cylinder bore (Chapter 6).
2   Lever balky or stuck. Check pivot and lubricate (Chapter 6).
3   Wheel cylinder piston seized in bore. Caused by wear or ingestion of dirt past deteriorated seal (Chapter 6).
4   Brake shoes damaged. Lining material separated from shoes. Usually caused by faulty manufacturing process or from contact with chemicals. Replace shoes (Chapter 6).
5   Shoes improperly installed (Chapter 6).
6   Rear brake pedal or lever free play insufficient (Chapter 1).
7   Rear brake springs weak. Replace brake springs (Chapter 6).

Electrical problems

53   Battery dead or weak

1   Battery faulty. Caused by sulfated plates which are shorted through sedimentation or low electrolyte level. Also, broken battery terminal making only occasional contact (Chapter 8).
2   Battery cables making poor contact (Chapter 8).
3   Load excessive. Caused by addition of high wattage lights or other electrical accessories.
4   Ignition switch defective. Switch either grounds/earths internally or fails to shut off system. Replace the switch (Chapter 8).
5   Regulator/rectifier defective (Chapter 8).
6   Stator coil open or shorted (Chapter 8).
7   Wiring faulty. Wiring grounded or connections loose in ignition, charging or lighting circuits (Chapter 8).

54   Battery overcharged

1   Regulator/rectifier defective. Overcharging is noticed when battery gets excessively warm or boils over (Chapter 8).
2   Battery defective. Replace battery with a new one (Chapter 8).
3   Battery amperage too low, wrong type or size. Install manufacturer's specified amp-hour battery to handle charging load (Chapter 8).

Chapter 1
Tune-up and routine maintenance

Contents

	Section		Section
Air cleaner - filter element and drain tube cleaning	14	Idle speed - check and adjustment	20
Battery - check	4	Introduction to tune-up and routine maintenance	2
Brake lever and pedal freeplay - check and adjustment	6	Lubrication - general	12
Brake system - general check	5	Reverse lock system - check and adjustment	8
Choke - operation check	11	Routine maintenance schedule	1
Clutch - check and freeplay adjustment	9	Spark plug - replacement	17
Cylinder compression - check	18	Steering system - inspection and toe-in adjustment	23
Engine oil/filter, differential oil and transfer case oil - change	13	Suspension - check	22
Exhaust system - inspection and spark arrester cleaning	16	Throttle operation/lever freeplay - check	
Fasteners - check	21	and adjustment	10
Fluid levels - check	3	Tires/wheels - general check	7
Fuel system - check and filter cleaning	15	Valve clearances - check and adjustment	19

Specifications

Engine

Spark plugs

Type
Standard	NGK DPR8EA-9 or ND X24EPR-U9
Extended high speed riding	NGK DPR9EA-9 or ND X27EPR-U9
Cold climates (below 5-degrees C/41-degrees F)	NGK DPR7EA-9 or ND X22EPR-U9
Gap	0.8 to 0.9 mm (0.031 to 0.035 inch)

Engine idle speed
1988 and 1990 models	1500 +/- 100 rpm
1991 and later models	1400 +/- 100 rpm
Valve clearance (COLD engine)	0.15 mm (0.006 inch)

Miscellaneous

Front brake shoe lining thickness
New	4 mm (0.16 inch)
Wear limit	1 mm (0.04 inch)

Rear brake shoe lining thickness
New	5 mm (0.20 inch)
Wear limit	2 mm (0.08 inch)
Front brake lever freeplay	25 to 30 mm (1 to 1-1/4 inches)
Rear brake pedal freeplay	15 to 20 mm (5/8 to 3/4 inch)
Rear brake lever freeplay	15 to 20 mm (5/8 to 3/4 inch)
Reverse selector lever freeplay	2 to 4 mm (1/16 to 1/8 inch)
Throttle lever freeplay	3 to 8 mm (1/8 to 5/16 inch)
Choke freeplay	Not adjustable
Minimum tire tread depth	4 mm (0.16 inch)

Miscellaneous (continued)

Tire pressures (cold)
  2WD models (front and rear tires)
    Standard .... 2.9 psi
    Maximum .... 3.3 psi
    Minimum .... 2.5 psi
  4WD models
    Front tires
      Standard... 4.4 psi
      Maximum.. 5.0 psi
      Minimum... 3.8 psi
    Rear tires
      Standard... 2.9 psi
      Maximum.. 3.3 psi
      Minimum... 2.5 psi
Front wheel toe-in
  2WD models
    1988 through 1992... 2 mm (5/64 inch)
    1993-on... 5 mm (13/64 inch)
  4WD models
    1988, 1990 .... 8 mm (5/64 inch)
    1991, 1992 .... 0 mm (0 inch)
    1993 on .... 4 mm (5/32 inch)

Torque specifications

Oil drain plug .... 25 Nm (18 ft-lbs)
Oil filter cover bolts.. 10 Nm (7 ft-lbs)
Clutch adjusting screw locknut .... 22 Nm(16 ft-lbs)
Valve adjusting screw locknuts .... 17 Nm (12 ft-lbs)
Spark plugs .... 18 Nm (13 ft-lbs)
Differential and transfer case filler plugs .... 12 Nm (108 in-lbs)
Differential drain bolts.. 12 Nm (108 in-lbs)
Transfer case check bolt .... not specified
Transfer case drain bolt .... 22 Nm (16 ft-lbs)
Tie-rod locknuts... 55 Nm (40 ft-lbs)

Recommended lubricants and fluids

Engine/transmission oil

Type.. API grade SF or SG multi-grade oil
Viscosity 10W-40
Capacity
  With filter change .... 2.25 liters (2.38 US qt, 3.96 Imp pt)
  Oil change only.. 2.2 liters (2.3 US qt, 3.8 Imp pt)
  After engine overhaul .... 2.5 liters (2.6 US qt, 4.4 Imp pt)

Transfer case oil

Type.. Engine oil, API grade SF or SG
Viscosity 10W-40
Capacity
  Oil change .... 190 cc (6.8 US fl oz)
  After overhaul .... 200 cc (6.4 US fl oz)

Differential oil

Type.. Hypoid gear oil
Viscosity .... SAE 80
Capacity
  Front differential
    Oil change... 190 cc (6.8 US fl oz)
    After overhaul.. 200 cc (6.4 US fl oz)
  Rear differential
    Oil change... 90 cc (3.0 US fl oz)
    After overhaul.. 100 cc (3.4 US fl oz)
Brake fluid... DOT 3 or DOT 4

Miscellaneous

Wheel bearings.. Medium weight, lithium-based multi-purpose grease (NLGI no. 3)
Swingarm pivot bearings.. Medium weight, lithium-based multi-purpose grease (NLGI no. 3)
Cables and lever pivots.. Chain and cable lubricant or 10W30 motor oil
Brake pedal/shift lever/throttle lever pivots............................. Chain and cable lubricant or 10W30 motor oil

1 Honda TRX300
Routine maintenance schedule

Note: *The pre-ride inspection outlined in the owner's manual covers checks and maintenance that should be carried out on a daily basis. It's condensed and included here to remind you of its importance. Always perform the pre-ride inspection at every maintenance interval (in addition to the procedures listed). The intervals listed below are the shortest intervals recommended by the manufacturer for each particular operation during the model years covered in this manual. Your owner's manual may have different intervals for your model.*

Daily or before riding

Check the engine oil level
Check the fuel level and inspect for leaks
Check the operation of both brakes - check the front brake fluid level and look for leakage; check the rear brake pedal and lever for correct freeplay
Check the tires for damage, the presence of foreign objects and correct air pressure
Check the throttle for smooth operation and correct freeplay
Make sure the steering operates smoothly
Check for proper operation of the headlight, taillight, indicator lights, speedometer and horn
Make sure the engine kill switch works properly
Check the driveaxle boots for damage or deterioration
Check the air cleaner drain tube and clean it if necessary
Make sure any cargo is properly loaded and securely fastened
Check all fasteners, including wheel nuts and axle nuts, for tightness
Check the underbody for mud or debris that could start a fire or interfere with vehicle operation

Every 100 operating hours or 600 miles

Perform all of the daily checks plus:
Check front brake fluid level

Inspect the brakes
Check and adjust the valve clearances
Clean the air filter element (1)
Clean the air cleaner housing drain tube (2)
Check/adjust the idle speed
Change the engine oil and filter
Inspect the suspension
Clean and gap the spark plugs
Check/adjust the reverse selector cable freeplay
Check the skid plates for looseness or damage
Adjust the clutch
Check the exhaust system for leaks and check fastener tightness; clean the spark arrester
Inspect the wheels and tires

Every 200 operating hours or 1200 miles

Perform all of the above checks plus:
Check/adjust the throttle lever freeplay
Check choke operation
Check the tightness of all fasteners
Check the cleanliness of the fuel system and the condition of the fuel line
Check the fuel tap strainer screen
Check the brake shoes for wear
Check the differential oil level
Check the transfer case oil level
Inspect the steering system and steering shaft bearing

Every two years

Change the brake fluid
Change the differential oil
Change the transfer case oil (4WD)

1   *More often in dusty, sandy or snowy conditions.*
2   *More often in wet or muddy conditions.*

2.1a Decals on the vehicle include maintenance information such as tire pressures . . .

2.1b . . . as well as safety information

2   Introduction to tune-up and routine maintenance

Refer to illustrations 2.1a and 2.1b

This Chapter covers in detail the checks and procedures necessary for the tune-up and routine maintenance of your vehicle. Section 1 includes the routine maintenance schedule, which is designed to keep the machine in proper running condition and prevent possible problems. The remaining Sections contain detailed procedures for carrying out the items listed on the maintenance schedule, as well as additional maintenance information designed to increase reliability. Maintenance information is also printed on decals, which are mounted in various locations on the vehicle **(see illustrations)**. Where information on the decals differs from that presented in this Chapter, use the decal information.

Since routine maintenance plays such an important role in the safe and efficient operation of your vehicle, it is presented here as a comprehensive checklist. For the rider who does all his own maintenance, these lists outline the procedures and checks that should be done on a routine basis.

Deciding where to start or plug into the routine maintenance schedule depends on several factors. If you have a vehicle whose warranty has recently expired, and if it has been maintained according to the warranty standards, you may want to pick up routine maintenance as it coincides with the next mileage or calendar interval. If you have owned the machine for some time but have never performed any maintenance on it, then you may want to start at the nearest interval and include some additional procedures to ensure that nothing important is overlooked. If you have just had a major engine overhaul, then you may want to start the maintenance routine from the beginning. If you have a used machine and have no knowledge of its history or maintenance record, you may desire to combine all the checks into one large service initially and then settle into the maintenance schedule prescribed.

The Sections which actually outline the inspection and maintenance procedures are written as step-by-step comprehensive guides to the actual performance of the work. They explain in detail each of the routine inspections and maintenance procedures on the check list. References to additional information in applicable Chapters is also included and should not be overlooked.

Before beginning any actual maintenance or repair, the machine should be cleaned thoroughly, especially around the oil filter housing, spark plugs, cylinder head covers, side covers, carburetor, etc. Cleaning will help ensure that dirt does not contaminate the engine and will allow you to detect wear and damage that could otherwise easily go unnoticed.

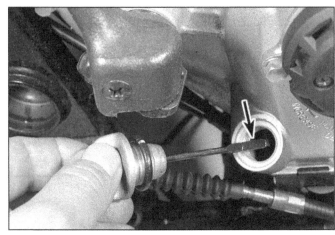

3.3 The engine oil level must be between the upper and lower marks on the dipstick

3   Fluid levels - check

Engine oil

Refer to illustration 3.3

1   Park the vehicle on a level surface, then start the engine and allow it to reach normal operating temperature. **Caution:** *Do not run the engine in an enclosed space such as a garage or shop.*

2   Stop the engine and allow the machine to sit undisturbed for about five minutes.

3   With the engine off, unscrew the dipstick from the right side of the crankcase. Pull it out, wipe it off with a clean rag, and reinsert it (let the dipstick rest on the threads; don't screw it back in). Pull the dipstick out and check the oil level on the dipstick scale. The oil level should be between the Maximum and Minimum level marks on the scale **(see illustration)**.

4   If the level is below the Minimum mark, add oil through the dipstick hole. Add enough oil of the recommended grade and type to bring the level up to the Maximum mark. Do not overfill.

Brake fluid

Refer to illustration 3.7

5   In order to ensure proper operation of the hydraulic front drum brakes, the fluid level in the master cylinder reservoirs must be properly maintained.

6   With the vehicle parked on a level surface, turn the handlebars until the top of the front brake master cylinder is as level as possible.

3.7 The brake fluid level must be above the LOWER mark on the reservoir; remove the cover screws (arrows) to add fluid

3.15a On 4WD models, unscrew the filler plug (upper arrow) to check the front differential oil level; unscrew the drain bolt (lower arrow) to change the oil

7   The fluid level is visible through the master cylinder reservoir. Make sure the fluid level is above the Lower mark on the reservoir **(see illustration)**.
8   If the level is low, the fluid must be replenished. Before removing the master cylinder cap, place rags beneath the reservoir (to protect the paint from brake fluid spills) and remove all dust and dirt from the area around the cap.
9   Remove the cover screws, then lift off the cover, rubber diaphragm and float (if equipped). **Caution:** *Don't operate the brake lever with the cover removed.*
10   Add new, clean brake fluid of the recommended type until the level is even with the cast line inside the master cylinder reservoir. Don't mix different brands of brake fluid in the reservoir, as they may not be compatible. Also, don't mix different specifications (DOT 3 with DOT 4).
11   Reinstall the float (if equipped), rubber diaphragm and cover. Tighten the cover screws to the torque listed in this Chapter's Specifications.
12   Wipe any spilled fluid off the reservoir body.
13   If the brake fluid level was low in either check, inspect the front or rear brake system for leaks.

Differential oil

Refer to illustrations 3.15a and 3.15b
14   Park the vehicle on a level surface.
15   Remove the differential filler cap **(see illustrations)**. Feel the oil level inside the differential; it should be up to the bottom of the filler threads.
16   Add oil if necessary of the type recommended in this Chapter's Specifications.
17   Reinstall the filler cap and tighten securely.

Transfer case oil (4WD models)

Refer to illustrations 3.18 and 3.19
18   To check the oil, remove the check bolt from the front of the transfer case **(see illustration)**. Oil should flow out of the hole.
19   If no oil flows, remove the filler plug **(see illustration)**.
20   Add oil of the type recommended in this Chapter's Specifications until it flows out of the check bolt hole. After the oil stops flowing, install the check bolt and filler plug and tighten them securely, but don't overtighten.

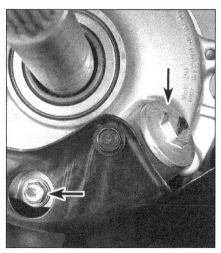

3.15b On all models, unscrew the filler plug (upper arrow) to check the rear differential oil level; unscrew the drain bolt (lower arrow) to change the oil

3.18 Remove the check bolt (upper arrow) to check the transfer case oil level on 4WD models; remove the drain bolt (lower arrow) to change the oil

3.19 Remove the filler plug at the top of the transfer case to add oil

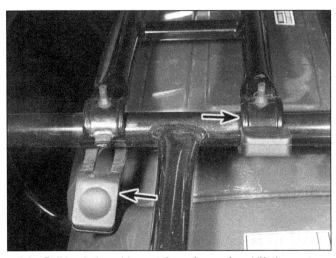

4.1a Pull back the rubber retainers (arrows) and lift the center portion of the rear rack . . .

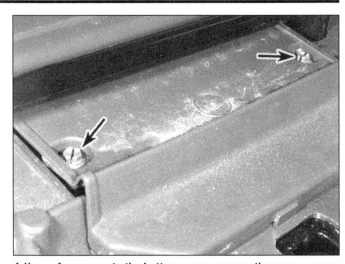

4.1b . . . for access to the battery cover; remove the cover screws (arrows) and lift the cover off

4.2a Disconnect the battery cables (negative cable first), then remove the battery retainer bolts (arrows) . . .

4.2b . . . and lift the battery out of the case

4   Battery - check

Refer to illustrations 4.1a, 4.1b, 4.2a and 4.2b

Warning: *Be extremely careful when handling or working around the battery. The electrolyte is very caustic and an explosive gas (hydrogen) is given off when the battery is charging. Always wear eye protection when working around the battery. Rinse off spilled electrolyte immediately with large amounts of water.*

1   Lift up the center section of the rear carrier and remove the battery cover **(see illustrations)**.

2   Remove the screws securing the battery cables to the battery terminals (remove the negative cable first, positive cable last). Undo the battery retaining clamp bolts and remove the battery **(see illustrations)**.

3   The battery is a sealed type which requires no maintenance. **Note:** *Do not attempt to remove the battery caps to check the electrolyte level or battery specific gravity. Removal will damage the caps, resulting in electrolyte leakage and battery damage. All that should be done is to check that its terminals are clean and tight and that the casing is not damaged or leaking. See Chapter 8 for further details.*

4   If the vehicle will be stored for an extended time, fully charge the battery, then disconnect the negative cable before storage.

5   Install the battery. Be sure to refer to safety precautions regarding battery installation in Chapter 8.

5   Brake system - general check

1   A routine general check of the brakes will ensure that any problems are discovered and remedied before the rider's safety is jeopardized.

2   Check the brake levers and pedal for loose connections, excessive play, bends, and other damage. Replace any damaged parts with new ones (see Chapter 6).

3   Make sure all brake fasteners are tight. Check the brake for wear as described below and make sure the fluid level in the reservoir is correct (see Section 3). Look for leaks at the hose connections and check for cracks in the hoses. If the lever is spongy, bleed the brakes as described in Chapter 6.

4   Make sure the brake light operates when the front brake lever is depressed. The front brake light switch is not adjustable. If it fails to operate properly, replace it with a new one (see Chapter 8).

5   Operate the rear brake lever and pedal. If operation is rough or sticky, refer to Section 12 and lubricate the cables.

Front brakes

6   Remove the adjusting hole plug from the brake drum (see Chapter 6).

7   Look through the hole to inspect the thickness of the lining

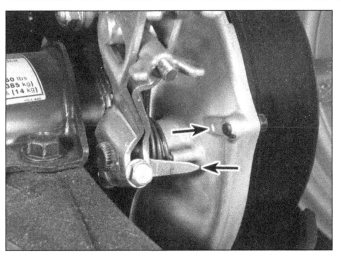

5.8 If the pointer (lower arrow) aligns with the mark (upper arrow) when the rear brake is applied, it's time to replace the rear brake shoes

6.1 Measure the freeplay of the front brake lever at the lever tip

6.3 Turn the adjuster wheel (arrow) to adjust the front brakes; 4WD models (shown) have an adjuster on each wheel cylinder (brake drum removed for clarity)

6.8 Loosen the lockwheel (right arrow) and turn the adjuster (left arrow) to adjust freeplay

material on the brake shoes (use a flashlight if necessary). If it's worn to near the limit listed in this Chapter's Specifications, refer to Chapter 6 and replace the brake shoes.

Rear brakes

Refer to illustration 5.8

8   With the rear brake lever and pedal freeplay properly adjusted (see Section 6), check the wear indicator on the rear brake panel **(see illustration)**. If the pointer lines up with the indicator when the lever is pulled or the pedal is pressed, refer to Chapter 6 and replace the brake shoes.

---

6   Brake lever and pedal freeplay - check and adjustment

Front brake lever

Refer to illustrations 6.1 and 6.3

1   Squeeze the front brake lever and note how far the lever travels **(see illustration)**. If it exceeds the limit listed in this Chapter's Specifications, adjust the front brakes as described below.

2   Securely block the rear wheels so the vehicle can't roll. Jack up the front end and support it securely on jackstands.

3   2WD models have two adjuster wheels, located on each side of the adjuster body at the bottom of the brake panel (see Chapter 6). 4WD models have an adjuster wheel at each wheel cylinder **(see illustration)**.

4   To adjust the brakes, insert a screwdriver through the adjusting hole and turn the adjuster wheel until the wheel can't be turned by hand, then back it off three notches. Spin the wheel by hand to make sure the brake lining isn't dragging on the drum; if it is, back off the adjuster just enough so the dragging stops. Then align the hole with the second adjuster wheel and repeat the adjustment.

5   Push the adjusting hole cap securely into its hole with a screwdriver.

6   Repeat the adjustment on the other front wheel, then remove the jackstands and lower the vehicle.

Rear brakes

Refer to illustrations 6.8, 6.9 and 6.10

7   Check the rear brake lever play at the left handlebar in the same way as for front brake lever play **(see illustration 6.1)**. If it exceeds the limit listed in this Chapter's Specifications, adjust it as described below.

8   Pull back the rubber cover from the handlebar adjuster **(see illustration)**. Loosen the lockwheel, turn the adjuster as needed to set freeplay, then tighten the lockwheel.

6.9   The upper wingnut adjusts the lever; the lower wingnut adjusts the pedal

6.10   Measure freeplay at the pedal

7.4   Check tire pressure with a gauge that will read accurately at the low pressures used in ATV tires

9   If freeplay can't be brought within specifications at the lockwheel adjuster, turn the lower wingnut at the brake panel lever **(see illustration)**. **Note:** *Push the lever forward so its bushing clears the cutout in the wingnut, then turn the nut. Make sure the cutout seats on the bushing after adjustment.*

10   Check the play of the rear brake pedal **(see illustration)**. If it exceeds the limit listed in this Chapter's Specifications, adjust it with the upper wingnut at the brake panel **(see illustration 6.9)**.

7   Tires/wheels - general check

Refer to illustration 7.4

1   Routine tire and wheel checks should be made with the realization that your safety depends to a great extent on their condition.

2   Check the tires carefully for cuts, tears, embedded nails or other sharp objects and excessive wear. Operation of the vehicle with excessively worn tires is extremely hazardous, as traction and handling are directly affected. Measure the tread depth at the center of the tire and replace worn tires with new ones when the tread depth is less than that listed in this Chapter's Specifications.

3   Repair or replace punctured tires as soon as damage is noted. Do not try to patch a torn tire, as wheel balance and tire reliability may be impaired.

4   Check the tire pressures when the tires are cold and keep them properly inflated **(see illustration)**. Proper air pressure will increase tire

life and provide maximum stability and ride comfort. Keep in mind that low tire pressures may cause the tire to slip on the rim or come off, while high tire pressures will cause abnormal tread wear and unsafe handling.

5   The steel wheels used on this machine are virtually maintenance free, but they should be kept clean and checked periodically for cracks, bending and rust. Never attempt to repair damaged wheels; they must be replaced with new ones.

6   Check the valve stem locknuts to make sure they're tight. Also, make sure the valve stem cap is in place and tight. If it is missing, install a new one made of metal or hard plastic.

8   Reverse lock system - check and adjustment

Refer to illustrations 8.2 and 8.3

1   Follow the reverse selector cable from its lever on the left handlebar to its lever on the right side of the engine. Check for kinks, bends, loose retainers or other problems and correct them as necessary.

2   Measure the gap between the reverse lock lever and its cable bracket at the handlebar **(see illustration)**. If it isn't within the range listed in this Chapter's Specifications, adjust it.

3   To adjust the cable, loosen the locknut at the right side of the engine **(see illustration)**. Turn the adjusting nut to achieve the correct play at the handlebar, then tighten the locknut securely.

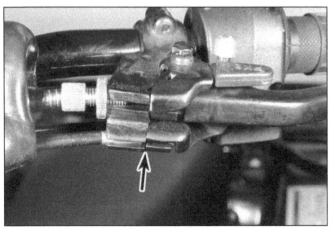

8.2   Measure freeplay at the gap between the reverse lock lever and its bracket (arrow)

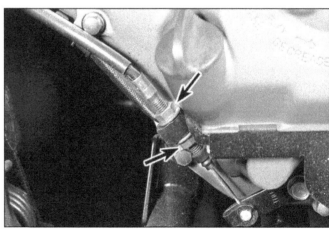

8.3   Loosen the locknut (lower arrow) and turn the adjusting nut (upper arrow) to adjust freeplay

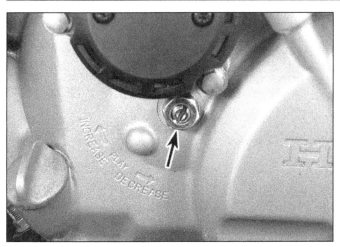

9.2 Loosen the locknut (arrow) and turn the screw as described in the text, then tighten the locknut

10.3 Loosen the lockwheel (right arrow) and turn the adjuster wheel (left arrow) to make fine adjustments in throttle freeplay

9 Clutch - check and freeplay adjustment

Refer to illustration 9.2

1   The automatic clutch mechanism on these models disengages the change clutch automatically when the shift lever is operated, so there is no clutch lever. If shifting gears becomes difficult, the clutch may be in need of adjustment.

2   Loosen the locknut on the right side of the engine **(see illustration)**. Carefully turn the adjusting screw counterclockwise (anti-clockwise) until you feel resistance, then turn it back in 1/4-turn. Hold the screw in this position and tighten the locknut to the torque listed in this Chapter's Specifications.

10 Throttle operation/lever freeplay - check and adjustment

Throttle check

1   Make sure the throttle lever moves easily from fully closed to fully open with the front wheel turned at various angles. The grip should return automatically from fully open to fully closed when released. If the throttle sticks, check the throttle cable for cracks or kinks in the housings. Also, make sure the inner cable is clean and well-lubricated.

2   Check for a small amount of freeplay at the lever and compare the freeplay to the value listed in this Chapter's Specifications.

Throttle adjustment

Refer to illustration 10.3

3   Freeplay adjustments can be made at the throttle lever end of the accelerator cable. Loosen the lockwheel on the cable **(see illustration)** and turn the adjuster until the desired freeplay is obtained, then retighten the lockwheel.

4   If the freeplay can't be adjusted at the grip end, adjust the cable at the carburetor end. To do this, first remove the fuel tank (see Chapter 3).

5   Loosen the locknut on the throttle cable (see Chapter 3). Turn the adjusting nut to set freeplay, then tighten the locknut securely.

11 Choke - operation check

1   Operate the choke lever on the left handlebar while you feel for smooth operation. If the lever doesn't move smoothly, refer to Section 12 and lubricate the choke cable.

2   Follow the cable from the handlebar to the starting enrichment valve on the engine **(see illustration 10.3a in Chapter 3)**. Check for kinks, bends, loose retainers or other problems and correct them as necessary.

12 Lubrication - general

Refer to illustration 12.3

1   Since the controls, cables and various other components of an ATV are exposed to the elements, they should be lubricated periodically to ensure safe and trouble-free operation.

2   The throttle and brake levers, brake pedal and kickstarter pivot should be lubricated frequently. In order for the lubricant to be applied where it will do the most good, the component should be disassembled. However, if chain and cable lubricant is being used, it can be applied to the pivot joint gaps and will usually work its way into the areas where friction occurs. If motor oil or light grease is being used, apply it sparingly as it may attract dirt (which could cause the controls to bind or wear at an accelerated rate). **Note:** *One of the best lubricants for the control lever pivots is a dry-film lubricant (available from many sources by different names).*

3   The throttle, choke, brake and reverse cables should be removed and treated with a commercially available cable lubricant which is specially formulated for use on vehicle control cables. Small adapters for pressure lubricating the cables with spray can lubricants are available and ensure that the cable is lubricated along its entire length **(see illustration)**. When attaching the cable to the lever, be sure to lubricate the barrel-shaped fitting at the end with multi-purpose grease.

4   To lubricate the cables, disconnect them at the lower end, then lubricate the cable with a pressure lube adapter **(see illustration 12.3)**.

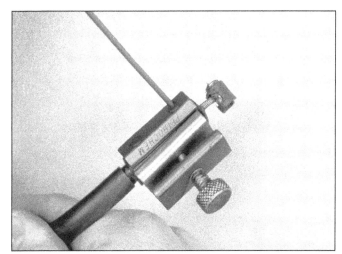

12.3 Lubricating a cable with a pressure lube adapter (make sure the tool seats around the inner cable)

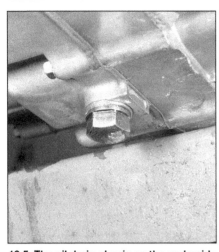

13.5 The oil drain plug is on the underside of the crankcase (skid plate removed for clarity)

13.6a Remove the filter cover bolts (arrows) and lift off the cover

13.6b Pull out the filter element; be sure its OUT-SIDE mark (arrow) faces outward on installation

See Chapter 2 for the reverse selector cable, Chapter 3 for the throttle and choke cables or Chapter 6 for the brake cables.

5   Refer to Chapter 5 for the following lubrication procedures:

  a) *Swingarm bearing and dust seals*
  b) *Front driveaxle splines (4WD models)*
  c) *Rear driveshaft pinion joint*
  d) *Rear axle shaft splines*

6   Refer to Chapter 6 for the following lubrication procedures:

  a) *Front wheel hub seals (2WD models)*
  b) *Brake pedal pivot and seals*
  c) *Front brake drum waterproof seals*
  d) *Rear brake drum cover seal*

---

13 Engine oil/filter, differential oil and transfer case oil - change

---

Engine oil/filter

Refer to illustrations 13.5, 13.6a, 13.6b, 13.6c, 13.10 and 13.12

1   Consistent routine oil and filter changes are the single most important maintenance procedure you can perform on a vehicle. The oil not only lubricates the internal parts of the engine, transmission and clutch, but it also acts as a coolant, a cleaner, a sealant, and a protectant. Because of these demands, the oil takes a terrific amount of abuse and should be replaced often with new oil of the recommended grade and type. Saving a little money on the difference in cost between a good oil and a cheap oil won't pay off if the engine is damaged. Honda recommends against using the following:

  a) *Oils with graphite or molybdenum additives*
  b) *Non-detergent oils*
  c) *Castor or vegetable based oils*
  d) *Oil additives.*

2   Before changing the oil and filter, warm up the engine so the oil will drain easily. Be careful when draining the oil, as the exhaust pipe, the engine and the oil itself can cause severe burns.

3   Park the vehicle over a clean drain pan.

4   Remove the dipstick/oil filler cap to vent the crankcase and act as a reminder that there is no oil in the engine.

5   Next, remove the drain plug from the engine **(see illustration)** and allow the oil to drain into the pan. Do not lose the sealing washer on the drain plug.

6   As the oil is draining, remove the oil filter cover bolts, then remove the cover, filter element and spring **(see illustrations)**. If additional maintenance is planned for this time period, check or service another component while the is allowed to drain completely.

13.6c Remove the spring from its post and wipe the remaining out of the filter housing; be sure to install a new O-ring (arrow) on installation

7   Wipe any remaining oil out of the filter housing area of the crankcase.

8   Check the condition of the drain plug threads and the sealing washer.

9   Install the spring on its post, then install the filter element **(see illustration 13.6c and 13.6b)**. **Caution:** *The OUT-SIDE mark on the filter must face outward (away from the crankcase) or severe engine damage will occur.*

10   Install new O-rings in the groove in the crankcase and on the filter cover **(see illustration 13.6c and the accompanying illustration)**. Install the cover and tighten its bolts to the torque listed in this Chapter's Specifications.

11   Slip a new sealing washer over the drain plug, then install and tighten the plug to the torque listed in this Chapter's Specifications. Avoid overtightening, as damage to the engine case will result.

12   Refer to Chapter 2 and remove the right side (clutch) cover from the engine. Pull out the filter screen **(see illustration)**, clean it with solvent and blow it dry. Install the new filter screen, then refer to Chapter 2 and install the right side cover.

13   Before refilling the engine, check the old oil carefully. If the oil was drained into a clean pan, small pieces of metal or other material can be easily detected. If the oil is very metallic colored, then the engine is

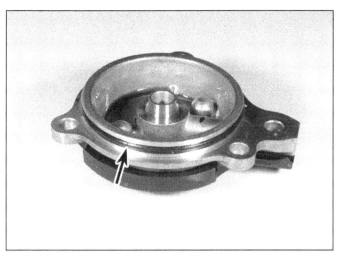

13.10 Install a new O-ring in the filter cover groove (arrow)

13.12 Pull the filter screen out of its slot, clean it and reinsert it

experiencing wear from break-in (new engine) or from insufficient lubrication. If there are flakes or chips of metal in the oil, then something is drastically wrong internally and the engine will have to be disassembled for inspection and repair.
14   If there are pieces of fiber-like material in the oil, the change clutch is experiencing excessive wear and should be checked.
15   If the inspection of the oil turns up nothing unusual, refill the crankcase to the proper level with the recommended oil and install the dipstick/filler cap. Start the engine and let it run for two or three minutes. Shut it off, wait a few minutes, then check the oil level. If necessary, add more oil to bring the level up to the upper level mark on the dipstick. Check around the drain plug and filter cover for leaks.
16   The old oil drained from the engine cannot be reused in its present state and should be disposed of. Check with your local refuse disposal company, disposal facility or environmental agency to see whether they will accept the oil for recycling. Don't pour used oil into drains or onto the ground. After the oil has cooled, it can be drained into a suitable container (capped plastic jugs, topped bottles, milk cartons, etc.) for transport to one of these disposal sites.

Differential oil

17   If you're working on a 4WD model, place a protective cover under the drain bolt so oil won't drip onto the frame rail. This can be a piece of cardboard or aluminum foil.
18   Place a drain pan beneath the differential.
19   Remove the oil filler plug, then the drain bolt and sealing washer **(see illustrations 3.15a or 3.15b).** Let the oil drain for several minutes, until it stops dripping.
20   Clean the drain bolt and sealing washer. If the sealing washer is in good condition, it can be reused; otherwise, replace it.
21   Install the drain bolt and tighten it to the torque listed in this Chapter's Specifications.
22   Add oil of the type and amount listed in this Chapter's Specifications, then install the filler plug and tighten it to the torque listed in this Chapter's Specifications.
23   Refer to Step 16 above to dispose of the drained oil.

Transfer case oil

24   Place a drain pan beneath the transfer case.
25   Remove the oil filler plug, then the drain bolt, check bolt and their sealing washers **(see illustrations 3.18 and 3.19).** Let the oil drain for several minutes, until it stops dripping.
26   Clean the drain bolt, check bolt and sealing washers. If the sealing washers are in good condition, they can be reused; otherwise, replace them.

27   Install the drain bolt and tighten it to the torque listed in this Chapter's Specifications.
28   Add oil of the type and amount listed in this Chapter's Specifications. **Note:** *Use the recommended capacity as a guide only. Transfer case oil is at the correct level when it starts to run out of the check bolt hole.*
29   Let the oil finish draining from the check bolt hole, then install the check bolt and its sealing washer. Tighten the bolt securely, but don't overtighten.
30   Install the filler plug and tighten it to the torque listed in this Chapter's Specifications.
31   Refer to Step 16 above to dispose of the drained oil.

14 Air cleaner - filter element and drain tube cleaning

Element cleaning
Refer to illustrations 14.2, 14.3 and 14.4
1   Remove the seat (see Chapter 7).
2   Remove the clips that secure the filter cover and lift it off **(see illustration).**

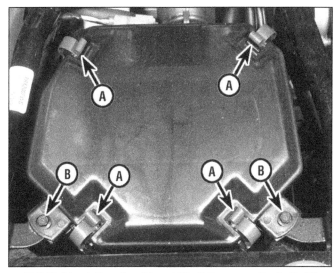

14.2 Pull back the clips (A) and lift off the cover; the bolts (B) need not be removed unless you plan to remove the air cleaner housing

14.3 Loosen the clamping band and remove the holder screw (arrows); then separate the filter element from the housing and lift out the holder, element and core

14.9 Open the clip and pull the drain tube (arrow) off its fitting

15.1 Check the fuel line for cracks, damage or deterioration and replace it if there are any problems

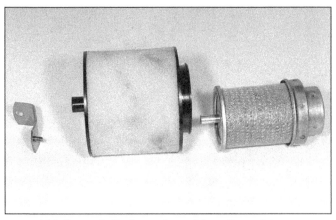

14.4 Separate the foam element from the metal core

3   Loosen the clamping band at the front of the element, remove the holder screw and lift the element out **(see illustration)**.
4   Separate the foam element from the metal core **(see illustration)**.
5   Clean the element and core in a high flash point solvent, squeeze the solvent out of the foam and let the core and element dry completely.
6   Soak the foam element in foam filter oil, then squeeze it firmly to remove the excess oil. Place the element on the core.
7   If you're working on a 1998 or later California model, check the separator for the crankcase ventilation system, located inside the air cleaner housing.
8   The remainder of installation is the reverse of the removal steps.

Drain tube cleaning

Refer to illustration 14.9
9   Check the drain tube for accumulated water and oil **(see illustration)**. If oil or water has built up in the tube, squeeze its clamp, remove it from the air cleaner housing and clean it out. Install the drain tube on the housing and secure it with the clamp.

---

15 Fuel system - check and filter cleaning

Refer to illustrations 15.1, 15.5a, 15.5b and 15.9
Warning: *Gasoline (petrol) is extremely flammable, so take extra precautions when you work on any part of the fuel system. Don't smoke or allow open flames or bare light bulbs near the work area, and don't work in a garage where a natural gas-type appliance (such as a water heater or clothes dryer) with a pilot light is present. Since gasoline is carcinogenic, wear latex gloves when there's a possibility of being exposed to fuel, and, if you spill any fuel on your skin, rinse it off immediately with soap and water. Mop up any spills immediately and do not store fuel-soaked rags where they could ignite. When you perform any kind of work on the fuel system, wear safety glasses and have a Class B type fire extinguisher on hand.*
1   Check the carburetor, fuel tank, the fuel tap and the line for leaks and evidence of damage **(see illustration)**.
2   If carburetor gaskets are leaking, the carburetor should be disassembled and rebuilt by referring to Chapter 3.
3   If the fuel tap is leaking, tightening the screws may help. If leakage persists, the tap should be disassembled and repaired or replaced with a new one.
4   If the fuel line is cracked or otherwise deteriorated, replace it with a new one.
5   Place the fuel tap lever in the Off position **(see illustration 15.1)**. Place a wrench on the hex at the bottom of the fuel tap and remove the cup, strainer and O-ring **(see illustrations)**.
6   Clean the strainer. If it's heavily clogged, remove the fuel tank (see Chapter 3). Unscrew the fuel valve nut **(see illustration 15.5a)**, then remove the fuel valve and clean the in-tank strainer.
7   Installation is the reverse of the removal steps, with the following additions:

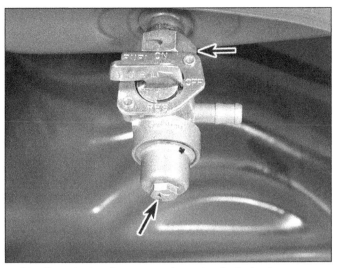

15.5a Turn the hex (lower arrow) counterclockwise to unscrew the cup from the fuel tap; drain the fuel tank and unscrew the fuel tap nut (upper arrow) if it's necessary to remove the in-tank strainer

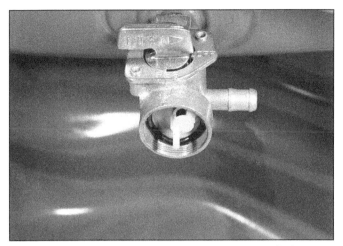

15.5b Pull the O-ring and plastic strainer from the fuel tap

a) *If the in-tank strainer was removed, install a new O-ring and tighten the nut to the torque listed in this Chapter's Specifications.*

b) *Use a new O-ring on the fuel tap cup. Hand-tighten the cup firmly, but don't overtighten or the O-ring will be squashed, resulting in fuel leaks.*

8   After installation, run the engine and check for fuel leaks.

9   If the vehicle will be stored for a month or more, remove and drain the fuel tank. Also loosen the float chamber drain screw and drain the fuel from the carburetor **(see illustration)**.

16   Exhaust system - inspection and spark arrester cleaning

Refer to illustration 16.3

1   Periodically check the exhaust system for leaks and loose fasteners. If tightening the holder nuts at the cylinder head fails to stop

any leaks, replace the gasket with a new one (a procedure which requires removal of the system).

2   The exhaust pipe flange nuts at the cylinder head are especially prone to loosening, which could cause damage to the head. Check them frequently and keep them tight.

3   **Warning:** *Make sure the exhaust system is cool before doing this procedure.* At the specified interval, remove the plate or plug from under the rear end of the muffler **(see illustration)**. Hold a wadded-up rag firmly over the hole at the rear end of the muffler. Have an assistant start the engine and rev it a few times to blow out carbon, then shut the engine off.

4   After the exhaust system has cooled, install the plate and gasket (1988 through 1991 models) or plug (1992 and later models). Tighten the bolts or plug securely.

17   Spark plug - replacement

Refer to illustrations 17.2a, 17.2b, 17.6a and 17.6b

1   This vehicle is equipped with a spark plug that has a 16 mm wrench hex.

2   Twist the spark plug cap to break it free from the plug, then pull it

15.9 If the vehicle will be stored more than a month, remove the float chamber drain screw (arrow) and its O-ring to drain the fuel

16.3 Unbolt the plate (early models) or unscrew the plug (later models, arrow), then hold a rag over the muffler opening and rev the engine a few times to blow carbon out of the spark arrester

17.2a Twist the spark plug cap back and forth to free it,
then pull it off the plug

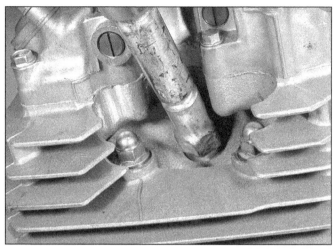

17.2b Unscrew the plug with a spark plug socket

off **(see illustration)**. If available, use compressed air to blow any accumulated debris from around the spark plug. Remove the plug **(see illustration)**.

3   Inspect the electrodes for wear. Both the center and side electrodes should have square edges and the side electrode should be of uniform thickness. Look for excessive deposits and evidence of a cracked or chipped insulator around the center electrode. Compare your spark plugs to the color spark plug reading chart. Check the threads, the washer and the ceramic insulator body for cracks and other damage.

4   If the electrodes are not excessively worn, and if the deposits can be easily removed with a wire brush, the plug can be regapped and reused (if no cracks or chips are visible in the insulator). If in doubt concerning the condition of the plug, replace it with a new one, as the expense is minimal.

5   Cleaning the spark plug by sandblasting is permitted, provided you clean the plug with a high flash-point solvent afterwards.

6   Before installing a new plug, make sure it is the correct type and heat range. Check the gap between the electrodes, as it is not preset. For best results, use a wire-type gauge rather than a flat gauge to check the gap **(see illustration)**. If the gap must be adjusted, bend the side electrode only and be very careful not to chip or crack the

insulator nose **(see illustration)**. Make sure the washer is in place before installing the plug.

7   Since the cylinder head is made of aluminum, which is soft and easily damaged, thread the plug into the head by hand. Slip a short length of hose over the end of the plug to use as a tool to thread it into place. The hose will grip the plug well enough to turn it, but will start to slip if the plug begins to cross-thread in the hole - this will prevent damaged threads and the accompanying repair costs.

8   Once the plug is finger tight, the job can be finished with a socket. If a torque wrench is available, tighten the spark plug to the torque listed in this Chapter's Specifications. If you do not have a torque wrench, tighten the plug finger tight (until the washer bottoms on the cylinder head) then use a wrench to tighten it an additional 1/4-turn. Regardless of the method used, do not over-tighten it.

9   Reconnect the spark plug cap.

---

18   Cylinder compression - check

Refer to illustration 18.5

1   Among other things, poor engine performance may be caused by

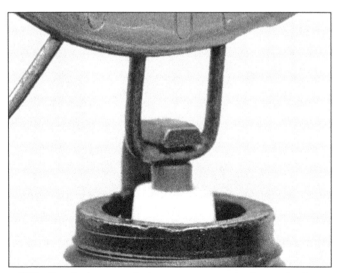

17.6a Spark plug manufacturers recommend using a wire type
gauge when checking the gap - if the wire doesn't slide between
the electrodes with a slight drag, adjustment is required

17.6b To change the gap, bend the side electrode only, as
indicated by the arrows, and be very careful not to crack or chip
the ceramic insulator surrounding
the center electrode

18.5 A compression gauge with a threaded fitting for the spark plug hole is preferred over the type that requires hand pressure to maintain the seal

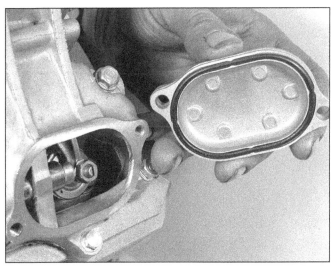

19.5 Unbolt the adjusting hole cover and lift it off; replace the O-ring if its condition is in doubt

leaking valves, incorrect valve clearances, a leaking head gasket, or worn piston, rings and/or cylinder wall. A cylinder compression check will help pinpoint these conditions and can also indicate the presence of excessive carbon deposits in the cylinder head.

2 The only tools required are a compression gauge and a spark plug wrench. Depending on the outcome of the initial test, a squirt-type oil can may also be needed.

3 Start the engine and allow it to reach normal operating temperature, then remove the spark plug (see Section 17, if necessary). Work carefully - don't strip the spark plug hole threads and don't burn your hands.

4 Disable the ignition by disconnecting the primary (low tension) wires from the coil (see Chapter 4). Be sure to mark the locations of the wires before detaching them.

5 Install the compression gauge in the spark plug hole **(see illustration)**. Hold or block the throttle wide open.

6 Crank the engine over a minimum of four or five revolutions (or until the gauge reading stops increasing) and observe the initial movement of the compression gauge needle as well as the final total gauge reading. Compare the results to the value listed in this Chapter's Specifications.

7 If the compression built up quickly and evenly to the specified amount, you can assume the engine upper end is in reasonably good mechanical condition. Worn or sticking piston rings and worn cylinders will produce very little initial movement of the gauge needle, but compression will tend to build up gradually as the engine spins over. Valve and valve seat leakage, or head gasket leakage, is indicated by low initial compression which does not tend to build up.

8 To further confirm your findings, add a small amount of engine oil to the cylinder by inserting the nozzle of a squirt-type oil can through the spark plug hole. The oil will tend to seal the piston rings if they are leaking.

9 If the compression increases significantly after the addition of the oil, the piston rings and/or cylinder are definitely worn. If the compression does not increase, the pressure is leaking past the valves or the head gasket. Leakage past the valves may be due to insufficient valve clearances, burned, warped or cracked valves or valve seats or valves that are hanging up in the guides.

10 If compression readings are considerably higher than specified, the combustion chamber is probably coated with excessive carbon deposits. It is possible (but not very likely) for carbon deposits to raise the compression enough to compensate for the effects of leakage past rings or valves. Refer to Chapter 2, remove the cylinder head and carefully decarbonize the combustion chamber.

19.6 Remove the timing hole cap (left arrow); on 4WD models, remove the reduction shaft hole cap (right arrow) and on 2WD models, remove the crankshaft hole cap

19 Valve clearances - check and adjustment

Refer to illustrations 19.5, 19.6, 19.7 and 19.9

1 The engine must be completely cool for this maintenance procedure (below 35-degrees C/95-degrees F), so if possible let the machine sit overnight before beginning.

2 Refer to Section 4 and disconnect the cable from the negative terminal of the battery.

3 Refer to Chapter 3 and remove the fuel tank.

4 Refer to Section 17 and remove the spark plug. This will make it easier to turn the engine.

5 Remove the valve adjusting hole covers (there's one on each side of the cylinder head **(see illustration)**.

6 Remove the timing hole cap. If you're working on a 4WD model, remove the reduction shaft cap **(see illustration)**. If you're working on a 2WD model, remove the crankshaft hole cap from the left side cover, just below the timing hole cap.

7 Position the piston at Top Dead Center (TDC) on the compression stroke. Do this by turning the crankshaft until the mark on the rotor is

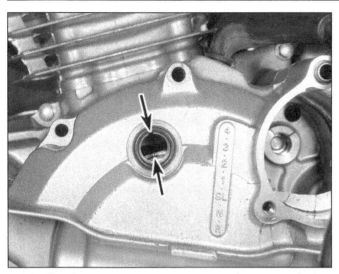

19.7 Align the notch on the crankcase cover (upper arrow) with the TDC line on the alternator rotor inside the hole (lower arrow) - the correct line has a sideways T to the left of it; the two lines next to each other are the advance timing mark, used to check ignition timing

aligned with the timing notch on the crankcase **(see illustration)**. If you're working on a 2WD model, turn the crankshaft clockwise with a socket on the alternator rotor bolt. If you're working on a 4WD model, turn the starter reduction shaft counterclockwise (anti-clockwise) with a 6 mm Allen wrench.

8   With the engine in this position, both of the valves can be checked.

9   To check, insert a feeler gauge of the thickness listed in this Chapter's Specifications between the valve stem and rocker arm **(see illustration)**. Pull the feeler gauge out slowly - you should feel a slight drag. If there's no drag, the clearance is too loose. If there's a heavy drag, the clearance is too tight.

10   If the clearance is incorrect, loosen the adjuster locknut with a box-end wrench (ring spanner). Turn the adjusting screw with a screwdriver until the correct clearance is achieved, then tighten the locknut. **Note:** *On some models, it's necessary to insert the screwdriver through an access hole in the frame to adjust the intake valve (on the rear side of the engine).*

11   After adjusting, recheck the clearance with the feeler gauge to make sure it wasn't changed when the locknut was tightened.

12   Now measure the other valve, following the same procedure you

19.9 Measure valve clearance with a feeler gauge; to adjust it, loosen the locknut and turn the adjusting screw with a screwdriver

used for the first valve. Make sure to use a feeler gauge of the specified thickness.

13   With both of the clearances within the Specifications, install the valve adjusting hole covers and timing hole caps. Install the crankshaft hole cap (2WD) or reduction shaft hole cap (4WD). Use new O-rings on the covers and caps if the old ones are hardened, deteriorated or damaged.

14   Install the fuel tank and reconnect the cable to the negative terminal of the battery.

---

20   Idle speed - check and adjustment

Refer to illustration 20.3

1   Before adjusting the idle speed, make sure the valve clearances and spark plug gap are correct. Also, turn the handlebars back-and-forth and see if the idle speed changes as this is done. If it does, the throttle cable may not be adjusted correctly, or it may be worn out. Be sure to correct this problem before proceeding.

2   The engine should be at normal operating temperature, which is usually reached after 10 to 15 minutes of stop and go riding. Make sure the transmission is in Neutral.

3   Turn the throttle stop screw **(see illustration)** until the idle speed

20.3 Turn the plastic knob on the throttle stop screw (arrow) to set the idle speed

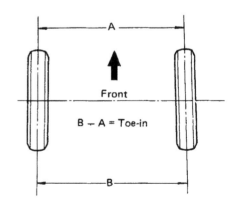

23.10 Toe-in measurement

listed in this Chapter's Specifications is obtained.

4   Snap the throttle open and shut a few times, then recheck the idle speed. If necessary, repeat the adjustment procedure.

5   If a smooth, steady idle can't be achieved, the fuel/air mixture may be incorrect. Refer to Chapter 4 for additional carburetor information.

21 Fasteners - check

1   Since vibration of the machine tends to loosen fasteners, all nuts, bolts, screws, etc. should be periodically checked for proper tightness. Also make sure all cotter pins or other safety fasteners are correctly installed.

2   Pay particular attention to the following:

Spark plug
Engine oil drain plug
Oil filter cover bolts
Gearshift lever
Brake pedal
Footpegs
Engine mount bolts
Shock absorber mount bolts
Front axle nuts
Rear axle nuts
Skid plate bolts

3   If a torque wrench is available, use it along with the torque specifications at the beginning of this, or other, Chapters.

22 Suspension - check

1   The suspension components must be maintained in top operating condition to ensure rider safety. Loose, worn or damaged suspension parts decrease the vehicle's stability and control.

2   Lock the front brake and push on the handlebars to compress the front shock absorbers several times. See if they move up-and-down smoothly without binding. If binding is felt, the shocks should be disassembled and inspected as described in Chapter 6.

3   Check the tightness of all front suspension nuts and bolts to be sure none have worked loose.

4   Inspect the rear shock absorber for fluid leakage and tightness of the mounting nuts and bolts. If leakage is found, the shock should be replaced.

5   Raise the rear of the vehicle and support it securely on jackstands. Grab the rear axle and rock the swingarm from side to side - there should be no discernible movement at the rear. If there's a little movement or a slight clicking can be heard, make sure the swingarm pivot shaft is tight. If the pivot shaft is tight but movement is still noticeable, the swingarm will have to be removed and the bearings replaced as described in Chapter 5.

6   Inspect the tightness of the rear suspension nuts and bolts.

23 Steering system - inspection and toe-in adjustment

Inspection

1   This vehicle is equipped with a ball bearing at the lower end of the steering shaft and a hard rubber bushing at the upper end, which can become dented, rough or loose during normal use of the machine. In extreme cases, worn or loose parts can cause steering wobble that is potentially dangerous.

2   To check the bearings, block the rear wheels so the vehicle can't roll, jack up the front end and support it securely on jackstands.

3   Point the wheel straight ahead and slowly move the handlebars from side-to-side. Dents or roughness in the bearing or bushing will be felt and the bars will not move smoothly. **Note:** *Make sure any hesitation in movement is not being caused by the cables and wiring harnesses that run to the handlebars.*

4   If the handlebars don't move smoothly, or if they move horizontally, refer to Chapter 5 to remove and inspect the steering shaft bushing and bearing.

Toe-in adjustment

Refer to illustrations 23.10 and 23.11

5   Remove the front carrier and fender (see Chapter 7).

6   Roll the vehicle forward onto a level surface and stop it with the front wheels pointing straight ahead.

7   Make a mark at the front and center of each tire, even with the centerline of the front hub.

8   Measure the distance between the marks with a toe-in gauge or steel tape measure.

9   Have an assistant push the vehicle backward while you watch the marks on the tires. Stop pushing when the tires have rotated exactly one-half turn, so the marks are at the backs of the tires.

10   Again, measure the distance between the marks. Subtract the front measurement from the rear measurement to get toe-in **(see illustration)**.

11   If toe-in is not as specified in this Chapter's Specifications, hold each tie-rod with a wrench on the flats and loosen the locknuts **(see illustration)**. Turn the tie-rods an equal amount to change toe-in. When toe-in is set correctly, tighten the locknuts to the torque listed in this Chapter's Specifications.

23.11 Hold and turn the tie rod by placing an open end wrench on the flats (A); loosen the locknuts (B) so the tie rod can be turned

Notes

Chapter 2
Engine, clutch and transmission

Contents

	Section
Cam chain tensioner - removal and installation	8
Camshaft and sprocket - removal, inspection and installation	9
Centrifugal clutch - removal, inspection and installation	17
Change clutch - removal, inspection and installation	18
Compression test	See Chapter 1
Crankcase - disassembly and reassembly	24
Crankcase components - inspection and servicing	25
Crankshaft and balancer - removal, inspection and installation	28
Cylinder - removal, inspection and installation	13
Cylinder head - removal and installation	10
Cylinder head and valves - disassembly, inspection and reassembly	12
Cylinder head cover and rocker arms - removal, inspection and installation	7
Engine - removal and installation	5
Engine disassembly and reassembly - general information	6
External oil pipe - removal and installation	16
External shift mechanism - removal, inspection and installation	23
General information	1

	Section
Initial start-up after overhaul	29
Kickstarter - removal, inspection and installation	22
Major engine repair - general note	4
Oil pipe and pump - removal, inspection and installation	20
Operations possible with the engine in the frame	2
Operations requiring engine removal	3
Output gear and countershaft - removal, inspection and installation	27
Piston - removal, inspection and installation	14
Piston rings - installation	15
Primary drive gear - removal, inspection and installation	21
Recommended break-in procedure	30
Reverse lock mechanism - cable replacement, removal, inspection and installation	19
Spark plug replacement	See Chapter 1
Transmission shafts and shift drum - removal, inspection and installation	26
Valves/valve seats/valve guides - servicing	11

Specifications

General

Bore	74 mm (2.91 inches)
Stroke	65.5 mm (2.58 inches)
Displacement	281.7 cc (17.2 cubic inches)

Rocker arms

Rocker arm inside diameter
Standard... 12.000 to 12.018 mm (0.4724 to 0.4731 inch)
Limit... 12.05 mm (0.474 inch)
Rocker shaft outside diameter
Standard... 11.966 to 11.984 mm (0.4711 to 0.4718 inch)
Limit... 11.92 mm (0.469 inch)
Shaft-to-arm clearance
Standard... 0.016 to 0.052 mm (0.0006 to 0.0020 inch)
Limit... 0.08 mm (0.003 inch)

Camshaft

Lobe height
    1988 through 1990 models
        Intake
            Standard ....   36.133 to 36.143 mm (1.4266 to 1.4229 inches)
            Limit ....   35.963 mm (1.4159 inches)
        Exhaust
            Standard ....   36.003 to 36.013 mm (1.4174 to 1.4178 inches)
            Limit ....   35.833 mm (1.4107 inches)
    1991 and later models
        Intake
            Standard ....   35.309 to 35.469 mm (1.3901 to 1.3964 inches)
            Limit ....   35.193 mm (1.3834 inches)
        Exhaust
            Standard ....   35.176 to 35.336 mm (1.3849 to 1.3912 inches)
            Limit ....   35.006 mm (1.3782 inches)
Bearing oil clearance
    Left and right journals
        Standard ....   0.025 to 0.067 mm (0.0010 to 0.0026 inch)
        Limit ....   0.10 mm (0.004 inch)
    Center journal
        Standard ....   0.045 to 0.087 mm (0.0018 to 0.0034 inch)
        Limit ....   0.12 mm (0.005 inch)
Journal diameter
    Right
        Standard ....   23.954 to 23.975 mm (0.9431 to 0.9439 inch)
        Limit ....   23.90 mm (0.941 inch)
    Center
        Standard ....   23.934 to 23.955 mm (0.9423 to 0.9431 inch)
        Limit ....   23.88 mm (0.940 inch)
    Left
        Standard ....   19.954 to 19.975 mm (0.7856 to 0.7864 inch)
        Limit ....   19.90 mm (0.783 inch)
Bearing journal inside diameter
    Right and center
        Standard ....   24.000 to 24.021 mm (0.9449 to 0.9457 inch)
        Limit ....   24.05 mm (0.947 inch)
    Left
        Standard ....   20.000 to 20.021 mm (0.7874 to 0.7882 inch)
        Limit ....   20.05 mm (0.789 inch)
Camshaft runout
    Standard ....   0.02 mm or less (0.0008 inch or less)
    Maximum ....   0.1 mm (0.0039 inch)

Cylinder head, valves and valve springs

Cylinder head warpage limit ....   0.10 mm (0.004 inch)
Valve stem runout ....   not specified
Valve stem diameter
    Intake
        Standard ....   5.475 to 5.490 mm (0.2156 to 0.2161 inch)
        Limit ....   5.45 mm (0.215 inch)
    Exhaust
        Standard ....   5.455 to 5.470 mm (0.2148 to 0.2154 inch)
        Limit ....   5.43 mm (0.214 inch)
Valve guide inside diameter (intake and exhaust)
    Standard ....   5.500 to 5.512 mm (0.2165 to 0.2170 inch)
    Limit ....   5.525 mm (0.2177 inch)
Valve seat width (intake and exhaust)
    Standard ....   1.2 mm (0.05 inch)
    Limit ....   1.5 mm (0.06 inch)
Valve spring free length
    Inner spring
        Standard ....   38.31 mm (1.508 inches)
        Limit ....   35.3 mm (1.39 inches)
    Outer spring
        Standard ....   46.83 mm (1.844 inches)
        Limit ....   43.8 mm (1.72 inches)

Cylinder

Bore diameter
    Standard.. 74.000 to 74.010 mm (2.9134 to 2.9138 inches)
    Limit.. 74.10 mm (2.917 inches)
Taper and out-of-round limits .... 0.10 mm (0.004 inch)

Pistons

Piston diameter
    Standard.. 73.960 to 73.985 mm (2.9118 to 2.9128 inches)
    Limit.. 73.90 mm (2.909 inches)
Piston-to-cylinder clearance
    Standard.. 0.15 to 0.50 mm (0.0006 to 0.0020 inch)
    Limit.. 0.10 mm (0.004 inch)
Oversize pistons and rings + 0.25 mm (0.010 inch), 0.5 mm (0.020 inch), 0.75 mm (0.030 inch), 1.00 mm (0.040 inch)
Piston pin bore
    In piston
        Standard .... 17.002 to 17.007 mm (0.6694 to 0.6696 inch)
        Limit .... 17.04 mm (0.671 inch)
    In connecting rod
        Standard .... 17.016 to 17.034 mm (0.6699 to 0.6706 inch)
        Limit .... 17.10 mm (0.673 inch)
Piston pin outer diameter
    Standard.. 16.994 to 17.000 mm (0.6691 to 0.6693 inch)
    Limit.. 16.96 mm (0.668 inch)
Piston pin-to-piston clearance
    Standard.. 0.002 to 0.014 mm (0.0001 to 0.0006 inch)
    Limit.. 0.02 mm (0.001 inch)
Ring side clearance
    Top
        Standard .... 0.02 to 0.05 mm (0.001 to 0.002 inch)
        Limit .... 0.09 mm (0.004 inch)
    Second
        Standard .... 0.015 to 0.045 mm (0.0006 to 0.0018 inch)
        Limit .... 0.09 mm (0.004 inch)
Ring end gap
    Top
        Standard .... 0.15 to 0.30 mm (0.006 to 0.012 inch)
        Limit .... 0.5 mm (0.02 inch)
    Second
        Standard .... 0.25 to 0.40 mm (0.010 to 0.016 inch)
        Limit .... 0.60 mm (0.02 inch)
    Oil
        Standard .... 0.20 to 0.70 mm (0.01 to 0.03 inch)
        Limit .... not specified

Centrifugal clutch

Drum internal diameter
    Standard.. 140.0 mm (5.51 inches)
    Limit.. 140.2 mm (5.52 inches)
Weight lining thickness
    Standard.. 3.0 mm (0.12 inch)
    Limit .... 2.0 mm (0.08 inch)
Clutch spring height
    Standard.. 3.1 mm (0.122 inch)
    Limit .... 2.95 mm (0.116 inch)
Clutch weight spring free length
    Standard.. 21.6 mm (0.85 inch)
    Limit .... 22.5 mm (0.89 inch)

Change clutch

Spring free length
    Standard.. 32.1 mm (1.26 inches)
    Limit.. 31.0 mm (1.22 inches)
Friction plate thickness
    Standard.. 2.62 to 2.78 mm (0.103 to 0.109 inch)
    Limit.. 2.3 mm (0.09 inch)
Friction and metal plate warpage limit .... 0.20 mm (0.008 inch)

Change clutch (continued)

Clutch outer guide outside diameter
    Standard... 27.959 to 27.980 mm (1.1007 to 1.1016 inches)
    Limit... 27.92 mm (1.099 inches)
Clutch outer guide inside diameter
    Standard... 22.000 t0 22.021 mm (0.8661 to 0.8670 inch)
    Limit... 22.05 mm (0.868 inch)
Diameter of outer guide friction surface on mainshaft
    Standard... 21.972 to 21.993 mm (0.8650 to 0.8659 inch)
    Limit... 21.93 mm (0.863 inch)

Primary drive gear

Inside diameter
    Standard... 27.000 to 27.012 mm (1.0630 to 1.0638 inches)
    Limit... 27.05 mm (1.065 inches)
Diameter of friction surface on crankshaft
    Standard... 26.959 to 26.980 mm (1.0614 to 1.0622 inches)
    Limit... 26.93 mm (1.060 inches)

Oil pump

Outer rotor to body clearance
    Standard... 0.15 to 0.21 mm (0.006 to 0.008 inch)
    Limit... 0.25 mm (0.010 inch)
Inner to outer rotor clearance
    Standard... 0.15 mm (0.006 inch) or less
    Limit... 0.20 mm (0.008 inch)
Side clearance (rotors to straightedge)
    Standard... 0.02 to 0.08 mm (0.001 to 0.003 inch)
    Limit... 0.10 mm (0.004 inch)

Kickstarter

Shaft outside diameter
    Standard... 23.959 to 23.980 mm (0.9433 to 0.9441 inch)
    Limit... 23.90 mm (0.941 inch)
Pinion gear inside diameter
    Standard... 24.000 to 24.021 mm (0.9449 to 0.9457 inch)

Shift drum and forks

Fork inside diameter
    Standard... 13.000 to 13.021 mm (0.5118 to 0.5126 inch)
    Limit... 13.04 mm (0.513 inch)
Fork shaft outside diameter
    Standard... 12.966 to 12.984 mm (0.5105 to 0.5112 inch)
    Limit... 12.96 mm (0.510 inch)
Fork ear thickness
    Standard... 4.93 to 5.00 mm (0.194 to 0.197 inch)
    Limit... 4.50 mm (0.177 inch)

Transmission

Gear inside diameters
    Mainshaft fourth
        Standard .... 25.000 to 25.021 mm (0.9843 to 0.9851 inch)
        Limit .... 25.05 mm (0.986 inch)
    Mainshaft fifth
        Standard .... 20.020 to 20.041 mm (0.7882 to 0.7890 inch)
        Limit .... 20.07 mm (0.790 inch)
    Countershaft first, second, third
        Standard .... 28.020 to 28.041 mm (1.1031 to 1.1040 inches)
        Limit .... 28.07 mm (1.105 inches)
    Countershaft reverse
        Standard .... 28.021 to 28.041 mm (1.1032 to 1.1040 inches)
        Limit .... 28.07 mm (1.105 inches)
    Reverse idler
        Standard .... 18.000 to 10.021 mm (0.7087 to 0.7095 inch)
        Limit .... 18.05 mm (0.711 inch)

Bushing diameters
    Mainshaft fourth outside diameter
        Standard ....   24.959 to 24.980 mm (0.9826 to 0.9835 inch)
        Limit ....   24.93 mm (0.981 inch)
    Mainshaft fourth inside diameter
        Standard ....   22.000 to 22.021 mm (0.8661 to 0.8670 inch)
        Limit ....   22.05 mm (0.868 inch)
    Mainshaft fifth outside diameter
        Standard ....   19.966 to 19.984 mm (0.7861 to 0.7868 inch)
        Limit ....   19.93 mm (0.785 mm)
    Mainshaft fifth inside diameter
        Standard ....   17.016 to 10.034 mm (0.6699 to 0.6706 inch)
        Limit ....   17.06 mm (0.672 inch)
    Countershaft first outside diameter
        Standard ....   27.984 to 28.005 mm (1.1017 to 1.1026 inches)
        Limit ....   27.93 mm (1.100 inches)
    Countershaft second/reverse outside diameter
        Standard ....   27.979 to 28.000 mm (1.1015 to 1.1024 inches)
        Limit ....   27.93 mm (1.100 inches)
    Countershaft third outside diameter
        Standard ....   27.984 to 28.005 mm (1.1017 to 1.1026 inches)
        Limit ....   27.93 mm (1.100 inches)
    Reverse outside diameter
        Standard ....   17.966 to 17.984 mm (0.7073 to 0.7080 inch)
        Limit ....   17.93 mm (0.706 inch)
    Reverse inside diameter
        Standard ....   14.000 to 14.025 mm (0.5512 to 0.5522 inch)
        Limit ....   14.05 mm (0.553 inch)
Gear-to-bushing clearances
    Mainshaft fourth
        Standard ....   0.020 to 0.062 mm (0.0008 to 0.0024 inch)
        Limit ....   0.10 mm (0.004 inch)
    Mainshaft fifth
        Standard ....   0.036 to 0.075 mm (0.0014 to 0.0030 inch)
        Limit ....   0.10 mm (0.004 inch)
    Countershaft first
        Standard ....   0.015 to 0.057 mm (0.0006 to 0.0022 inch)
        Limit ....   0.10 mm (0.004 inch)
    Countershaft second/reverse
        Standard ....   0.020 to 0.062 mm (0.0008 to 0.0024 inch)
        Limit ....   0.10 mm (0.00 inch)
    Countershaft third
        Standard ....   0.015 to 0.057 mm (0.0006 to 0.0022 inch)
        Limit ....   0.10 mm (0.004 inch)
    Reverse idler
        Standard ....   0.016 to 0.055 mm (0.0006 to 0.0022 inch)
        Limit ....   0.10 mm (0.004 inch)
Shaft diameters
    Mainshaft fourth gear surface
        Standard ....   21.959 to 21.980 mm (0.8645 to 0.8654 inch)
        Limit ....   21.93 mm (0.863 inch)
    Mainshaft fifth gear surface
        Standard ....   16.983 to 16.994 mm (0.6686 to 0.6691 inch)
        Limit ....   16.95 mm (0.667 inch)
    Reverse idler gear surface
        Standard ....   13.966 to 13.984 mm (0.5498 to 0.5506 inch)
        Limit ....   13.93 mm (0.548 inch)
Shaft-to-bushing clearances
    Mainshaft fourth
        Standard ....   0.020 to 0.062 mm (0.0008 to 0.0024 inch)
        Limit ....   0.10 mm (0.004 inch)
    Mainshaft fifth
        Standard ....   0.022 to 0.051 mm (0.0009 to 0.0020 inch)
        Limit ....   0.10 mm (0.004 inch)
    Reverse idler
        Standard ....   0.016 to 0.059 mm (0.0006 to 0.0023 inch)
        Limit ....   0.10 mm (0.004 inch)
Output gear backlash
    Standard..   0.080 to 0.180 mm (0.0031 to 0.0071 inch)
    Limit..   0.25 mm (0.010 inch)

Crankshaft

Connecting rod side clearance	
Standard..	0.05 to 0.65 mm (0.002 to 0.026 inch)
Limit...	0.80 mm (0.031 inch)
Connecting rod big end radial clearance	
Standard..	0.006 to 0.018 mm (0.0002 to 0.0007 inch)
Limit...	0.05 mm (0.002 inch)
Runout limit..	0.05 mm (0.002 inch)

Torque specifications

Bearing retainer plate (outside of right crankcase).................................	12 Nm (108 in-lbs) (3)
Cam chain guide cup bolt	12 Nm (108 in-lbs) (3)
Cam chain rear side guide bolt..	12 Nm (108 in-lbs) (3)
Cam chain tensioner	
Cap bolt...	10 Nm (84 in-lbs)
Mounting bolts	10 Nm (84 in-lbs)
Cam sprocket bolts	20 Nm (168 in-lbs)
Clutch	
Centrifugal clutch locknut	120 Nm (87 ft-lbs) (1)
Change clutch spring bolts	12 Nm (108 in-lbs)
Change clutch locknut	110 Nm (80 ft-lbs) (2)
Crankcase cover bolts (left or right)..	10 Nm (84 in-lbs)
Crankcase bolts...	10 Nm (84 in-lbs)
Cylinder base bolts..	10 Nm (84 in-lbs)
Cylinder head cover	
Flange bolts	12 Nm (108 in-lbs)
Standard bolts..	10 Nm (84 in-lbs)
Cylinder head	
Allen bolts..	25 Nm (18 ft-lbs)
Cap nuts...	40 Nm (29 ft-lbs)
Engine bracket bolts (on cylinder head and front of engine)	
1992 and earlier...	55 Nm (40 ft-lbs)
1993 on	75 Nm (54 ft-lbs)
Engine mounting bolt/nut (rear)	75 Nm (54 ft-lbs)
External oil pipe bolts	12 Nm (108 in-lbs)
External shift linkage return spring pin	22 Nm (16 ft-lbs) (3)
Internal oil pipe dark-colored bolt	12 Nm (108 in-lbs)
Kickstarter ratchet guide bolts	12 Nm (108 in-lbs)
Oil pump cover bolts (1994 on)	7 Nm (62 inch-lbs)
Output gear mounting bolts	32 Nm (23 ft-lbs)
Reverse/neutral rotor bolt...	12 Nm (108 in-lbs) (3)
Shift pedal pinch bolt..	16 Nm (144 in-lbs)

1   *Use non-permanent thread locking agent on the threads. Turn counterclockwise (anti-clockwise) to tighten. Stake the locknut.*
2   *Use non-permanent thread locking agent on the threads. Stake the locknut.*
3   *Use non-permanent thread locking agent on the threads.*

1   General information

The engine/transmission unit is of the air-cooled, single-cylinder four-stroke design. The two valves are operated by an overhead camshaft which is chain driven off the crankshaft. The engine/transmission assembly is constructed from aluminum alloy. The crankcase is divided vertically.

The crankcase incorporates a wet sump, pressure-fed lubrication system which uses a gear-driven rotor-type oil pump, an oil filter and separate strainer screen and an oil temperature warning switch.

Power from the crankshaft is routed to the transmission via two clutches. The centrifugal clutch, which engages as engine speed is increased, connects the crankshaft to the change clutch, which is of the wet, multi-plate type. The change clutch transmits power to the transmission; it's engaged and disengaged automatically when the shift lever is moved from one gear position to another. The transmission has five forward gears and one reverse gear.

2   Operations possible with the engine in the frame

The components and assemblies listed below can be removed without having to remove the engine from the frame. If, however, a number of areas require attention at the same time, removal of the engine is recommended.

> *Gear selector mechanism external components*
> *Kickstarter*
> *Starter motor*
> *Starter reduction gears*
> *Starter clutch*
> *Alternator rotor and stator*
> *Clutch assemblies*
> *Cam chain tensioner*
> *Camshaft*
> *Rocker arm assembly*
> *Cylinder head*
> *Cylinder and piston*
> *Oil pump*

3   Operations requiring engine removal

It is necessary to remove the engine/transmission assembly from the frame and separate the crankcase halves to gain access to the following components:

> *Crankshaft and connecting rod*
> *Transmission shafts*
> *Shift drum and forks*

4   Major engine repair - general note

1   It is not always easy to determine when or if an engine should be completely overhauled, as a number of factors must be considered.
2   High mileage is not necessarily an indication that an overhaul is needed, while low mileage, on the other hand, does not preclude the need for an overhaul. Frequency of servicing is probably the single most important consideration. An engine that has regular and frequent oil and filter changes, as well as other required maintenance, will most likely give many miles of reliable service. Conversely, a neglected engine, or one which has not been broken in properly, may require an overhaul very early in its life.
3   Exhaust smoke and excessive oil consumption are both indications that piston rings and/or valve guides are in need of attention. Make sure oil leaks are not responsible before deciding that the rings and guides are bad. Refer to Chapter 1 and perform a cylinder compression check to determine for certain the nature and extent of the work required.
4   If the engine is making obvious knocking or rumbling noises, the connecting rod and/or main bearings are probably at fault.
5   Loss of power, rough running, excessive valve train noise and high fuel consumption rates may also point to the need for an overhaul, especially if they are all present at the same time. If a complete tune-up does not remedy the situation, major mechanical work is the only solution.
6   An engine overhaul generally involves restoring the internal parts to the specifications of a new engine. During an overhaul the piston rings are replaced and the cylinder walls are bored and/or honed. If a rebore is done, then a new piston is also required. The crankshaft and connecting rod are permanently assembled, so if one of these components needs to be replaced both must be. Generally the valves are serviced as well, since they are usually in less than perfect condition at this point. While the engine is being overhauled, other components such as the carburetor and the starter motor can be rebuilt also. The end result should be a like-new engine that will give as many trouble-free miles as the original.
7   Before beginning the engine overhaul, read through all of the related procedures to familiarize yourself with the scope and requirements of the job. Overhauling an engine is not all that difficult, but it is time consuming. Plan on the vehicle being tied up for a minimum of two (2) weeks. Check on the availability of parts and make sure that any necessary special tools, equipment and supplies are

obtained in advance.
8   Most work can be done with typical shop hand tools, although a number of precision measuring tools are required for inspecting parts to determine if they must be replaced. Often a dealer service department or repair shop will handle the inspection of parts and offer advice concerning reconditioning and replacement. As a general rule, time is the primary cost of an overhaul so it doesn't pay to install worn or substandard parts.
9   As a final note, to ensure maximum life and minimum trouble from a rebuilt engine, everything must be assembled with care in a spotlessly clean environment.

5   Engine - removal and installation

Note: *Engine removal and installation should be done with the aid of an assistant to avoid damage or injury that could occur if the engine is dropped. A hydraulic floor jack should be used to support and lower the engine if possible (they can be rented at low cost).*

Removal

Refer to illustrations 5.10, 5.13a, 5.13b and 5.13c
1   Drain the engine and transfer case oil (see Chapter 1).
2   Disconnect the negative battery cable from the engine (it's secured by one of the starter mounting bolts) (see Chapter 8).
3   Remove the fuel tank, carburetor and exhaust system (see Chapter 3). The choke and throttle cables can be left connected. Plug the carburetor intake opening with a rag.
4   Disconnect the spark plug wire (see Chapter 1).
5   Remove the right footpeg (see Chapter 7) and the brake pedal (see Chapter 6).
6   Label and disconnect the following wires (refer to Chapters 4 or 8 for component location if necessary):

> *Ignition pulse generator*
> *Alternator*
> *Oil temperature, reverse and neutral switches*
> *Starter cable*

7   Disconnect the reverse cable (see Section 19).
8   Remove the shift pedal (see Section 23).
9   Loosen the swingarm boot clamps (see Chapter 5).
10   Disconnect the crankcase breather tube at the joint above the engine **(see illustration)**. Follow the breather tubes to the engine and remove any retainers that will obstruct engine removal.
11   If you're working on a 4WD model, remove the front drive side

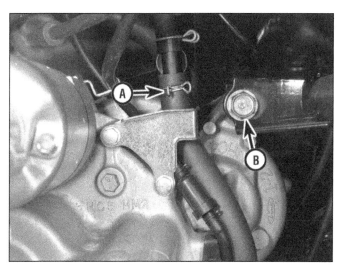

5.10 Disconnect the breather hose clamp

> *A   Breather hose clamp*
> *B   Upper rear engine mount bolt*

5.13a Remove the upper front engine mount through-bolt (A) and unbolt the bracket on each side (B)

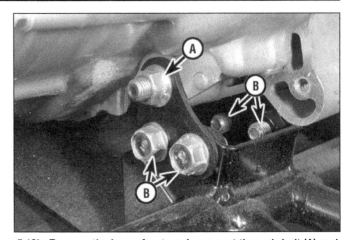

5.13b Remove the lower front engine mount through-bolt (A) and the bracket bolts on each side (B)

shaft and transfer case (see Chapter 5).

12   Support the engine securely from below.

13   Remove the engine mounting bolts, nuts and brackets at the upper front, upper rear, lower front and lower rear **(see illustration 5.10 and the accompanying illustrations)**.

14   Have an assistant help you lift the engine. As you lift, disengage the driveshaft from the output gear at the left rear corner of the engine. The driveshaft is mounted inside the left side of the swingarm (refer to Chapter 5 if necessary). Remove the engine to the right side of the vehicle.

15   Slowly lower the engine to a suitable work surface.

Installation

Refer to illustration 5.16

16   Check the rubber engine supports for wear or damage and replace them if necessary before installing the engine **(see illustration)**.

17   Coat the driveshaft splines with moly-based grease. Lift the engine up into its installed position in the frame and engage the driveshaft with the output gear (refer to Chapter 5 if necessary).

18   Place a floor jack under the engine at the jacking point just behind the lower front engine mount. Lift the engine to align the mounting bolt holes, then install the brackets, bolts and nuts. Tighten them to the torques listed in this Chapter's Specifications.

19   The remainder of installation is the reverse of the removal steps, with the following additions:

a) *Use new gaskets at all exhaust pipe connections.*

b) *Adjust the throttle cable and reverse selector cable following the procedures in Chapter 1.*

c) *Fill the engine and transfer case with oil, also following the procedures in Chapter 1. Run the engine and check for leaks.*

---

6   Engine disassembly and reassembly - general information

Refer to illustrations 6.2 and 6.3

1   Before disassembling the engine, clean the exterior with a degreaser and rinse it with water. A clean engine will make the job easier and prevent the possibility of getting dirt into the internal areas of the engine.

2   In addition to the precision measuring tools mentioned earlier, you will need a torque wrench, a valve spring compressor, oil gallery brushes **(see illustration)**, a piston ring removal and installation tool, a piston ring compressor and a clutch holder tool (which is described in Section 18). Some new, clean engine oil of the correct grade and type, some engine assembly lube (or moly-based grease) and a tube of RTV (silicone) sealant will also be required.

3   An engine support stand made from short lengths of 2 x 4's bolted together will facilitate the disassembly and reassembly procedures **(see illustration)**. If you have an automotive-type engine stand, an adapter plate can be made from a piece of plate, some angle iron and some nuts and bolts.

4   When disassembling the engine, keep "mated" parts together

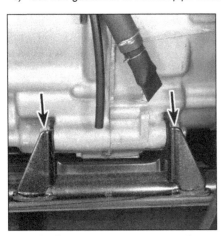

5.13c Remove the lower rear engine mount through-bolt and nut (arrows)

5.16 Replace the rubber engine supports if they're damaged or deteriorated

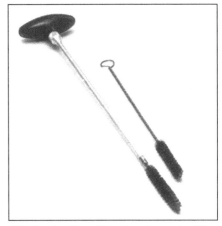

6.2 A selection of brushes is required for cleaning holes and passages in the engine components

6.3 An engine stand can be made from short lengths of lumber
and lag bolts or nails

(including gears, drum shifter pawls, etc.) that have been in contact
with each other during engine operation. These "mated" parts must be
reused or replaced as an assembly.
5   Engine/transmission disassembly should be done in the following
general order with reference to the appropriate Sections.
    Remove the cylinder head cover and rocker assembly
    Remove the cam chain tensioner and camshaft
    Remove the cylinder head
    Remove the cylinder
    Remove the piston
    Remove the clutches
    Remove the oil pump
    Remove the external shift mechanism
    Remove the alternator rotor
    Separate the crankcase halves
    Remove the shift drum/forks
    Remove the transmission shafts/gears
    Remove the crankshaft and connecting rod
6   Reassembly is accomplished by reversing the general
disassembly sequence.

7   Cylinder head cover and rocker arms - removal, inspection and installation

Note: *The cylinder head cover can be removed with the engine in the
frame. If the engine has been removed, ignore the steps which don't apply.*

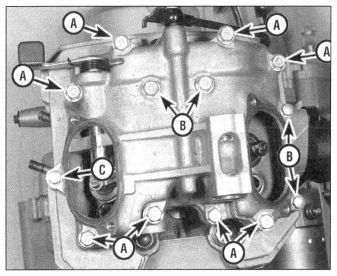

7.5 Loosen the cover bolts in two or three stages, in a
criss-cross pattern

A   *Flange head bolts*
B   *Flange head bolts (remove together with cover)*
C   *Standard bolts*

Removal

Refer to illustrations 7.5, 7.6 and 7.7
1   Remove the fuel tank (see Chapter 3).
2   Remove the external oil pipe from the engine (see Section 16).
3   Remove the front and rear upper engine mounting brackets (see
Section 5).
4   Refer to *Valve clearances - check and adjustment* in Chapter 1
and place the engine at top dead center on the compression stroke.
5   Loosen the cylinder head cover bolts in 2 or 3 stages, in a criss-
cross pattern **(see illustration)**.
6   Lift the cover off the cylinder head, together with two of the
cylinder head bolts **(see illustration 7.5 and the accompanying illustration)**. If it's stuck, don't attempt to pry it off - tap around the sides of
it with a plastic hammer to dislodge it.
7   Locate the cover dowels **(see illustration)**. They may be in the
cylinder head or they may have come off with the cover. Stuff clean rags
into the cam chain openings so small parts or tools can't fall into them.

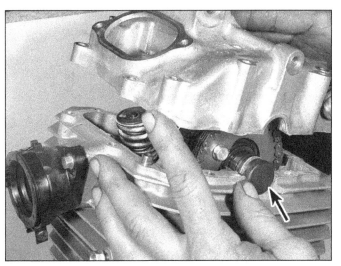

7.6 Lift off the cover and remove the camshaft end seal (arrow)

7.7 Locate the cover dowels (arrows)

Inspection

Refer to illustrations 7.8, 7.10 and 7.11

8   Check the rocker arms for wear at the cam contact surfaces and at the tips of the valve adjusting screws **(see illustration)**. Also check the decompression lug on the exhaust rocker arm for wear at the point where it contacts the decompression shaft. Try to twist the rocker arms from side-to-side on the shafts. If they're loose on the shafts or if there's visible wear, remove them as described below.

9   Grind a notch into the side of each dowel so the removal tool can grip it **(see illustration 7.8)**. Attach Honda pin puller 07936-MA70100 and slide hammer 07936-3710200 or equivalent and pull the dowels out of the cover.

10   Slide the rocker shafts out of the cover and remove the rocker arms **(see illustration)**.

11   Check the decompression lever and its shaft for wear, damage or a broken spring **(see illustration)**. If any problems are found, remove the locating bolt and pull the lever/shaft out of the cover, together with the spring. Pry the shaft oil seal out of its bore.

12   Measure the outer diameter of each rocker shaft and the inner diameter of the rocker arms with a micrometer and compare the measurements to the values listed in this Chapter's Specifications. If rocker arm-to-shaft clearance is excessive, replace the rocker arm or shaft, whichever is worn.

13   If the decompression lever/shaft was removed, press a new oil seal into the bore with a seal driver or a socket the same diameter as the seal. Install the shaft, engaging its spring with the cylinder head cover and the lever **(see illustration 7.11)**. Install the locating bolt, make sure it engages the shaft groove and tighten it securely.

14   Coat new rocker shaft O-rings with clean engine oil and install them on the shafts.

15   Coat the rocker shafts and rocker arm bores with moly-based grease. Install the rocker shafts and rocker arms in the cylinder head cover. The intake rocker arm can be identified by the "I" mark cast into it. The exhaust rocker shaft can be identified by its two holes. Be sure to install the intake and exhaust rocker arms and shafts in the correct side of the cover. As you install the rocker shafts, align their holes with the dowel holes and the cover mounting bolt holes.

16   Install new dowel pins in the cylinder head over, making sure they go all the way through the rocker shafts and seat in the cover.

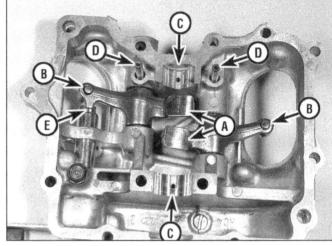

7.8 Cylinder head cover details

A Rocker arm cam contact surfaces	C Cam bearing surfaces
	D Rocker shaft dowels
B Valve adjusting screws	E Decompression lug

Installation

Refer to illustrations 7.19 and 7.20

17   Inspect the camshaft end seal **(see illustration 7.6)**. If the seal is cracked, hardened or deteriorated, pull it off and install a new one.

18   Clean the mating surfaces of the cylinder head and the valve cover with lacquer thinner, acetone or brake system cleaner. Apply a thin film of RTV sealant to the cover sealing surface (there's no gasket).

19   Install the two bolts in the cover **(see illustration 7.5)**. Position the cover on the cylinder head, making sure the camshaft end seal is in position **(see illustration)**.

20   Install a new copper washer on the cover bolt marked with an arrowhead **(see illustration)**. Install the bolts and tighten them evenly in two or three stages (starting with the center bolt) to the torque listed in this Chapter's Specifications. Note that there are two types of bolts

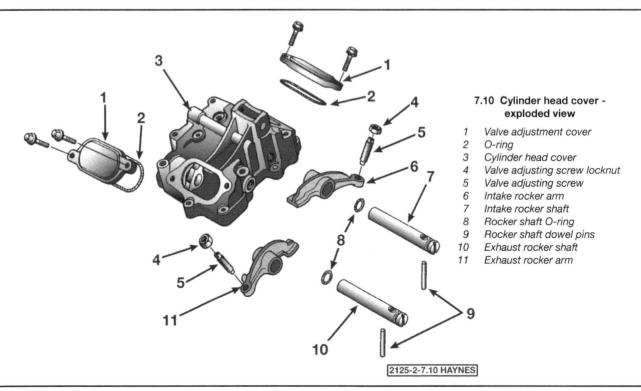

7.10 Cylinder head cover - exploded view

1   Valve adjustment cover
2   O-ring
3   Cylinder head cover
4   Valve adjusting screw locknut
5   Valve adjusting screw
6   Intake rocker arm
7   Intake rocker shaft
8   Rocker shaft O-ring
9   Rocker shaft dowel pins
10   Exhaust rocker shaft
11   Exhaust rocker arm

2125-2-7.10 HAYNES

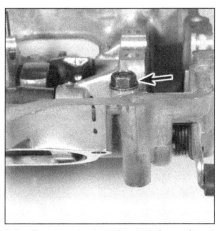

7.11 Remove the locating bolt (arrow) and pull the decompression shaft out of the cover

7.19 Make sure the cover secures the camshaft end seal

7.20 One cover bolt has a copper washer; it's marked with an arrowhead (arrow)

with different torque settings.
21   Adjust the valve clearances (see Chapter 1).
22   The remainder of installation is the reverse of removal.

8   Cam chain tensioner - removal and installation

Removal

Refer to illustration 8.1
1   Loosen the tensioner cap bolt **(see illustration)**.
2   Remove the tensioner mounting bolts and detach it from the cylinder block. **Caution:** *The tensioner piston locks in place as it extends. Once the tensioner bolts have been loosened, the tensioner piston must be reset before the bolts are tightened. If the bolts are loosened partway and then retightened without resetting the tensioner piston, the piston will be forced against the cam chain, damaging the tensioner or the chain.*
3   Remove the cap bolt and sealing washer from the tensioner body and wash them with solvent.

Installation

Refer to illustration 8.6
4   Clean all old gasket material from the tensioner body and engine.
5   Lubricate the friction surfaces of the components with moly-

based grease.
6   Turn the tensioner shaft clockwise with a screwdriver to retract the tensioner piston into the body **(see illustration)**. Insert a thin piece of stiff wire into one of the slots next to the cap bolt hole to lock the tensioner piston in position.
7   Install a new tensioner gasket on the cylinder. Position the tensioner body on the cylinder and install the bolts, tightening them to the torque listed in this Chapter's Specifications.
8   Pull the wire out of the cap bolt hole. Install the cap bolt with a new sealing washer and tighten it to the torque listed in this Chapter's Specifications.

9   Camshaft and sprocket - removal, inspection and installation

Note: *This procedure can be performed with the engine in the frame.*

Removal

Refer to illustrations 9.4, 9.5a and 9.5b
1   Refer to the valve adjustment procedure in Chapter 1 and position the piston at top dead center (TDC) on its compression stroke.
2   Remove the cylinder head cover following the procedure given in Section 7.
3   Remove the camshaft chain tensioner (see Section 8).

8.1 Loosen the tensioner cap bolt (arrow)

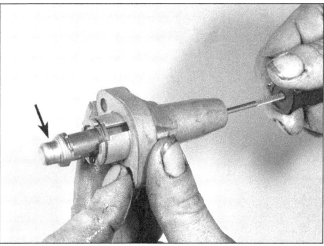

8.6 Insert a small screwdriver into the tensioner and turn it clockwise to retract the piston (arrow) (piston shown extended)

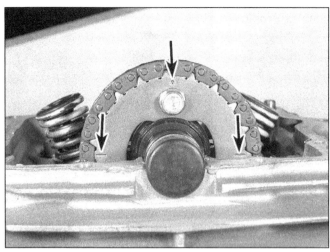

9.4 With the engine at TDC compression, the sprocket lines should be even with the cylinder head surface (lower arrows) and the punch mark should be straight up (upper arrow); remove the bolt next to the punch mark, then turn the crankshaft to expose the other sprocket bolt and remove it

9.5a Disengage the sprocket from the chain and lift out the camshaft and sprocket . . .

4   Remove one of the sprocket bolts **(see illustration)**. Turn the crankshaft clockwise one revolution to expose the other sprocket bolt, then remove it and pull off the camshaft end seal.

5   Pull up on the camshaft chain and carefully guide the camshaft out **(see illustration)**. With the chain still held taut, tie it to the engine with a piece of wire so it doesn't drop down off the sprocket **(see illustration)**. Engine damage could occur if the engine is rotated with the chain bunched up around the crank sprocket.

6   Pull the exhaust side (front) cam chain guide out of its slot. To remove the intake side (rear) cam chain guide, it's necessary to remove the clutch cover (see Section 17) so the bolt, washer and bushing can be removed **(see illustration 9.20b)**.

7   Cover the top of the cylinder head with a rag to prevent foreign objects from falling into the engine.

Inspection

Refer to illustrations 9.8, 9.9a and 9.9b

Note: *Before replacing camshafts or the cylinder head cover and cylinder head because of damage, check with local machine shops specializing in ATV or motorcycle engine work. In the case of the*

camshaft, it may be possible for cam lobes to be welded, reground and hardened, at a cost lower than that of a new camshaft. If the bearing surfaces in the cylinder head or cover are damaged, it may be possible for them to be bored out to accept bearing inserts. Due to the cost of a new cylinder head it is recommended that all options be explored before condemning it as trash!

8   Inspect the cam bearing surfaces of the head and the bearing caps **(see illustration 7.8 and the accompanying illustration)**. Look for score marks, deep scratches and evidence of spalling (a pitted appearance).

9   Check the camshaft lobes for heat discoloration (blue appearance), score marks, chipped areas, flat spots and spalling **(see illustration)**. Measure the height of each lobe with a micrometer **(see illustration)** and compare the results to the minimum lobe height listed in this Chapter's Specifications. If damage is noted or wear is excessive, the camshaft must be replaced. Also, be sure to check the condition of the rocker arms, as described in Section 7.

10   Except in cases of oil starvation, the camshaft chain wears very little. If the chain has stretched excessively, which makes it difficult to maintain proper tension, replace it with a new one. To remove the chain from the crankshaft sprocket, remove the centrifugal clutch (see Section 17).

11   Check the sprocket for wear, cracks and other damage, replacing it if necessary. If the sprocket is worn, the chain is also worn, and

9.5b . . . then tie the chain up so it doesn't drop down off the crankshaft sprocket

9.8 Check the cam bearing surfaces (arrows) for wear or damage

9.9a Check the cam lobes for wear - here's a good example of damage which will require replacement (or repair) of the camshaft

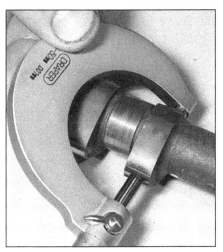

9.9b Measure the height of the cam lobes with a micrometer

9.14 Lubricate the journals with moly-based grease or assembly lube, then install the camshaft with the lobes downward and one bolt hole straight up

9.20a The pins on the front side chain guide fit in sockets in the cylinder (cylinder head removed for clarity)

9.20b The bottom ends of the cam chain guides should look like this when installed

possibly the sprocket on the crankshaft. If wear this severe is apparent, the entire engine should be disassembled for inspection.

12   Check the chain guides for wear or damage. If they are worn or damaged, replace them.

Installation

Refer to illustrations 9.14, 9.15, 9.20a and 9.20b

13   Make sure the bearing surfaces in the cylinder head and cylinder head cover are clean.

14   Lubricate the cam bearing journals with molybdenum disulfide grease. Lay the camshaft in the cylinder head with the lobes downward and one of the sprocket bolt holes straight up **(see illustration)**.

15   Engage the sprocket with the chain and align the sprocket bolt hole that has a punch mark next to it with the camshaft bolt hole **(see illustration 9.4)**.

16   Check the position of the timing marks on the sprocket. The lines should be even with the cylinder head cover gasket surface **(see illustration 9.4)**. The punch mark should be straight up. Recheck the crankshaft timing mark to make sure it's still at the TDC position (see *Valve clearance - check and adjustment* in Chapter 1).

17   Apply non-permanent thread locking agent to the threads of the sprocket bolt. Install it and tighten it to the torque listed in this

Chapter's Specifications.

18   Rotate the crankshaft to expose the other sprocket bolt hole. Coat the bolt threads with non-permanent thread locking agent, install the bolt and tighten it to the torque listed in this Chapter's Specifications.

19   Rotate the engine back to the TDC position and recheck the timing marks. With the crankshaft mark in the TDC position, the sprocket lines and punch mark should be aligned correctly **(see illustration 9.4)**. If this isn't the case, try removing the sprocket and repositioning it in the chain. Don't run the engine with the marks out of alignment or severe engine damage could occur.

20   Install the cam chain guides **(see illustrations)**.

21   Coat the cam lobes with moly-based grease or engine assembly lube.

22   Install the cylinder head cover (see Section 7).

23   Install the cam chain tensioner as described in Section 8.

24   Adjust the valve clearances (see Chapter 1).

25   The remainder of installation is the reverse of removal.

10 Cylinder head - removal and installation

Caution: *The engine must be completely cool before beginning this procedure, or the cylinder head may become warped.*

Removal

Refer to illustration 10.5

1   Remove the exhaust system. Loosen the clamp that secures the intake tube to the carburetor (see Chapter 3).

2   Remove the cylinder head cover, cam chain tensioner and camshaft (Sections 7, 8 and 9).

3   Loosen the cylinder head Allen bolts and cap nuts in two or three stages in the reverse of the tightening sequence **(see illustration 10.11)**. Remove the nuts, bolts and washers (use needle-nose pliers to remove the washers if necessary).

4   Lift the cylinder head off the cylinder. If the head is stuck, tap around the side of the head with a rubber mallet to jar it loose, or use two wooden dowels inserted into the intake or exhaust ports to lever the head off. Don't attempt to pry the head off by inserting a screwdriver between the head and the cylinder block - you'll damage the sealing surfaces.

5   Support the cam chain so it won't drop into the cam chain tunnel, and stuff a clean rag into the cam chain tunnel to prevent the entry of debris. Once this is done, remove the gasket and the two dowel pins

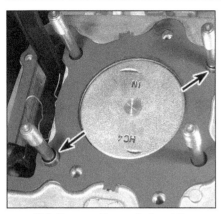

10.5 Remove the gasket and the dowels (arrows)

10.11 Cylinder head nut and bolt TIGHTENING sequence; the cap nuts have steel washers and the Allen bolts have copper washers

12.7a Install a valve spring compressor, compress the spring and remove the keepers (collets), then release the compressor and remove the valve spring retainer, springs, spring seat and valve

from the cylinder **(see illustration)**.

6   Check the cylinder head gasket and the mating surfaces on the cylinder head and cylinder for leakage, which could indicate warpage. Refer to Section 12 and check the flatness of the cylinder head.

7   Clean all traces of old gasket material from the cylinder head and cylinder. Be careful not to let any of the gasket material fall into the crankcase, the cylinder bore or the bolt holes.

Installation

Refer to illustration 10.11

8   Install the two dowel pins, then lay the new gasket in place on the cylinder block. Never re-use the old gasket and don't use any type of gasket sealant.

9   Carefully lower the cylinder head over the studs and dowels. It is helpful to have an assistant support the camshaft chain with a piece of wire so it doesn't fall and become kinked or detached from the crankshaft. When the head is resting on the cylinder, wire the cam chain to another component to keep tension on it.

10   Install the steel washers and cap nuts on the studs. Install new copper washers on the Allen bolts, then thread them light into their holes.

11   Using the proper sequence **(see illustration)**, tighten the nuts and bolts in two or three stages to the torque listed in this Chapter's Specifications.

12   Install all parts removed for access, making sure that the cam chain front guide engages the notch in the crankcase.

13   Change the engine oil (see Chapter 1).

11   Valves/valve seats/valve guides - servicing

1   Because of the complex nature of this job and the special tools and equipment required, servicing of the valves, the valve seats and the valve guides (commonly known as a valve job) is best left to a professional.

2   The home mechanic can, however, remove and disassemble the head, do the initial cleaning and inspection, then reassemble and deliver the head to a dealer service department or properly equipped repair shop for the actual valve servicing. Refer to Section 12 for those procedures.

3   The dealer service department will remove the valves and springs, recondition or replace the valves and valve seats, replace the valve guides, check and replace the valve springs, spring retainers and keepers (as necessary), replace the valve seals with new ones and reassemble the valve components.

4   After the valve job has been performed, the head will be in like-new condition. When the head is returned, be sure to clean it again very thoroughly before installation on the engine to remove any metal

particles or abrasive grit that may still be present from the valve servicing operations. Use compressed air, if available, to blow out all the holes and passages.

12   Cylinder head and valves - disassembly, inspection and reassembly

1   As mentioned in the previous Section, valve servicing and valve guide replacement should be left to a dealer service department or other repair shop. However, disassembly, cleaning and inspection of the valves and related components can be done (if the necessary special tools are available) by the home mechanic. This way no expense is incurred if the inspection reveals that service work is not required at this time.

2   To properly disassemble the valve components without the risk of damaging them, a valve spring compressor is absolutely necessary. If the special tool is not available, have a dealer service department or other repair shop handle the entire process of disassembly, inspection, service or repair (if required) and reassembly of the valves.

Disassembly

Refer to illustrations 12.7a, 12.7b and 12.7c

3   Remove the cylinder head (see Section 10).

4   Before the valves are removed, scrape away any traces of gasket material from the head gasket sealing surface. Work slowly and do not nick or gouge the soft aluminum of the head. Gasket removing solvents, which work very well, are available at most vehicle shops and auto parts stores.

5   Carefully scrape all carbon deposits out of the combustion chamber area. A hand held wire brush or a piece of fine emery cloth can be used once most of the deposits have been scraped away. Do not use a wire brush mounted in a drill motor, or one with extremely stiff bristles, as the head material is soft and may be eroded away or scratched by the wire brush.

6   Before proceeding, arrange to label and store the valves along with their related components so they can be kept separate and reinstalled in the same valve guides they are removed from (again, plastic bags work well for this).

7   Compress the valve spring(s) on the first valve with a spring compressor, then remove the keepers (collets) **(see illustration)**. Do not compress the spring(s) any more than is absolutely necessary. Carefully release the valve spring compressor and remove the retainer, spring(s), spring seat and valve from the head **(see illustration)**. If the valve binds in the guide (won't pull through), push it back into the head and deburr the area around the keeper (collet) groove with a very fine file or whetstone **(see illustration)**.

8   Repeat the procedure for the remaining valves. Remember to

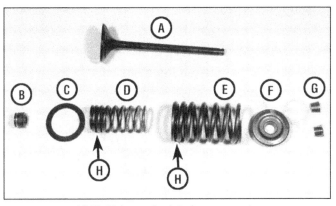

12.7b Valve and related components

A	Valve stem	E	Outer valve spring
B	Oil seal	F	Valve spring retainer
C	Spring seat	G	Keepers (collets)
D	Inner valve spring	H	Tightly wound coils

keep the parts for each valve together so they can be reinstalled in the same location.

9   Once the valves have been removed and labeled, pull off the valve stem seals with pliers and discard them (the old seals should never be re-used).

10   Next, clean the cylinder head with solvent and dry it thoroughly. Compressed air will speed the drying process and ensure that all holes and recessed areas are clean.

11   Clean all of the valve springs, keepers/collets, retainers and spring seats with solvent and dry them thoroughly. Do the parts from one valve at a time so that no mixing of parts between valves occurs.

12   Scrape off any deposits that may have accumulated on the valve, then use a motorized wire brush to remove deposits from the valve heads and stems. Again, make sure the valves do not get mixed up.

Inspection

Refer to illustrations 12.14a, 12.14b, 12.15, 12.16, 12.17, 12.18a, 12.18b, 12.19a and 12.19b

13   Inspect the head very carefully for cracks and other damage. If cracks are found, a new head will be required. Check the cam bearing surfaces for wear and evidence of seizure. Check the camshaft for wear as well (see Section 9).

14   Using a precision straightedge and a feeler gauge, check the head gasket mating surface for warpage. Lay the straightedge lengthwise, across the head and diagonally (corner-to-corner), intersecting the head bolt holes, and try to slip a feeler gauge under it, on either side of each combustion chamber **(see illustrations).** The

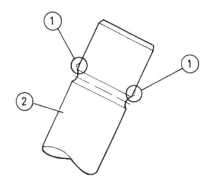

12.7c Check the area around the keeper (collet) groove for burrs and remove any that you find

1	Burrs (remove)	2	Valve stem

12.14a Check the gasket surface for flatness with a straightedge and feeler gauge . . .

feeler gauge thickness should be the same as the cylinder head warpage limit listed in this Chapter's Specifications. If the feeler gauge can be inserted between the head and the straightedge, the head is warped and must either be machined or, if warpage is excessive, replaced with a new one.

15   Examine the valve seats in each of the combustion chambers. If they are pitted, cracked or burned, the head will require valve service that is beyond the scope of the home mechanic. Measure the valve seat width **(see illustration)** and compare it to this Chapter's Specifications. If it is not within the specified range, or if it varies around its circumference, valve service work is required.

16   Clean the valve guides to remove any carbon buildup, then measure the inside diameters of the guides (at both ends and the

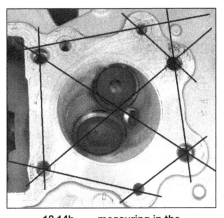

12.14b . . . measuring in the directions shown

12.15 Measuring valve seat width

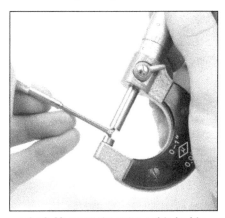

12.16 Measure the valve guide inside diameter with a hole gauge, then measure the gauge with a micrometer

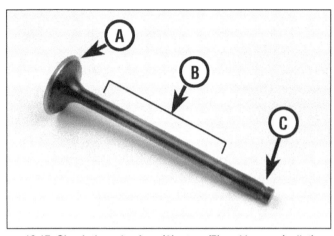

12.17 Check the valve face (A), stem (B) and keeper (collet) groove (C) for wear and damage

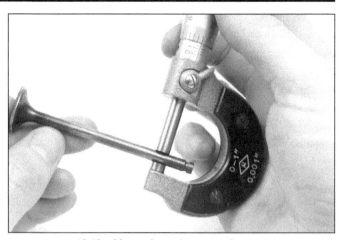

12.18a Measuring valve stem diameter

center of the guide) with a small hole gauge and a micrometer **(see illustration).** Record the measurements for future reference. The guides are measured at the ends and at the center to determine if they are worn in a bell-mouth pattern (more wear at the ends). If they are, guide replacement is an absolute must.

17   Carefully inspect each valve face for cracks, pits and burned spots. Check the valve stem and the keeper groove area for cracks **(see illustration).** Rotate the valve and check for any obvious indication that it is bent. Check the end of the stem for pitting and excessive wear and make sure the bevel is the specified width. The presence of any of the above conditions indicates the need for valve servicing.

18   Measure the valve stem diameter **(see illustration).** If the diameter is less than listed in this Chapter's Specifications, the valves will have to be replaced with new ones. Also check the valve stem for bending. Set the valve in a V-block with a dial indicator touching the middle of the stem **(see illustration).** Rotate the valve and look for a reading on the gauge (which indicates a bent stem). If the stem is bent, replace the valve.

19   Check the end of each valve spring for wear and pitting. Measure the free length **(see illustration)** and compare it to this Chapter's Specifications. Any springs that are shorter than specified have sagged and should not be reused. Stand the spring on a flat surface and check it for squareness **(see illustration).**

20   Check the spring retainers and keepers/collets for obvious wear and cracks. Any questionable parts should not be re-used, as extensive damage will occur in the event of failure during engine operation.

21   If the inspection indicates that no service work is required, the valve components can be reinstalled in the head.

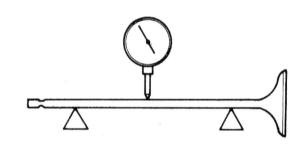

12.18b Check the valve stem for bends with a V-block (or V-blocks, as shown here) and a dial indicator

Reassembly

Refer to illustrations 12.23, 12.24a, 12.24b, 12.26 and 12.27

22   If the valve seats have been ground, the valves and seats should be lapped before installing the valves in the head to ensure a positive seal between the valves and seats. This procedure requires coarse and fine valve lapping compound (available at auto parts stores) and a valve lapping tool. If a lapping tool is not available, a piece of rubber or plastic hose can be slipped over the valve stem (after the valve has been installed in the guide) and used to turn the valve.

23   Apply a small amount of coarse lapping compound to the valve face **(see illustration),** then slip the valve into the guide. **Note:** *Make sure the valve is installed in the correct guide and be careful not to get any lapping compound on the valve stem.*

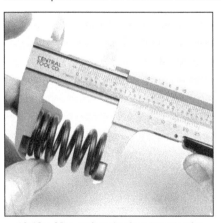

12.19a Measuring the free length of a valve spring

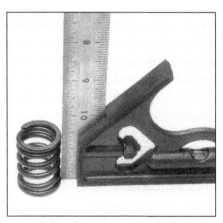

12.19b Checking a valve spring for squareness

12.23 Apply the lapping compound very sparingly, in small dabs, to the valve face only

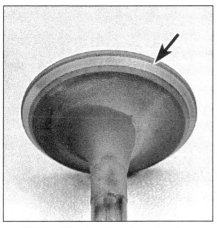

12.24a After lapping, the valve face should exhibit a uniform, unbroken contact pattern (arrow) . . .

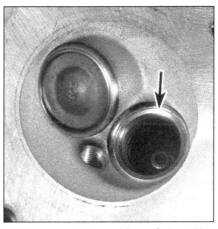

12.24b . . . and the seat (arrow) should be the specified width with a smooth, unbroken appearance

12.26 Push the oil seal onto the valve guide (arrow)

24   Attach the lapping tool (or hose) to the valve and rotate the tool between the palms of your hands. Use a back-and-forth motion rather than a circular motion. Lift the valve off the seat and turn it at regular intervals to distribute the lapping compound properly. Continue the lapping procedure until the valve face and seat contact area is of uniform width and unbroken around the entire circumference of the valve face and seat **(see illustrations)**. Once this is accomplished, wipe the valve and seat clean and lap the valves again with fine lapping compound.
25   Carefully remove the valve from the guide and wipe off all traces of lapping compound. Use solvent to clean the valve and wipe the seat area thoroughly with a solvent soaked cloth. Repeat the procedure for the remaining valves.
26   Lay the spring seat in place in the cylinder head, then install new valve stem seals on both of the guides **(see illustration)**. Use an appropriate size deep socket to push the seals into place until they are properly seated. Don't twist or cock them, or they will not seal properly against the valve stems. Also, don't remove them again or they will be damaged.
27   Coat the valve stems with assembly lube or moly-based grease, then install one of them into its guide. Next, install the spring seat, springs and retainers, compress the springs and install the keepers/collets. **Note:** *Install the springs with the tightly wound coils at the bottom (next to the spring seat). When compressing the springs with the valve spring compressor, depress them only as far as is absolutely necessary to slip the keepers/collets into place. Apply a small amount of grease to the keepers/collets* **(see illustration)** *to help hold them in place as the pressure is released from the springs. Make*

certain that the keepers/collets are securely locked in their retaining grooves.
28   Support the cylinder head on blocks so the valves can't contact the workbench top, then very gently tap each of the valve stems with a soft-faced hammer. This will help seat the keepers in their grooves.
29   Once all of the valves have been installed in the head, check for proper valve sealing by pouring a small amount of solvent into each of the valve ports. If the solvent leaks past the valve(s) into the combustion chamber area, disassemble the valve(s) and repeat the lapping procedure, then reinstall the valve(s) and repeat the check. Repeat the procedure until a satisfactory seal is obtained.

13   Cylinder - removal, inspection and installation

Removal

Refer to illustrations 13.3, 13.4 and 13.5
1   Following the procedure given in Section 10, remove the cylinder head. Make sure the crankshaft is positioned at Top Dead Center (TDC).
2   Lift out the cam chain front guide.
3   Remove two bolts that secure the base of the cylinder to the crankcase **(see illustration)**.
4   Lift the cylinder straight up to remove it **(see illustration)**. If it's stuck, tap around its perimeter with a soft-faced hammer (but don't tap on the cooling fins or they may break). Don't attempt to pry between the cylinder and the crankcase, as you'll ruin the sealing surfaces.

12.27 A small dab of grease will help hold the keepers/collets in place on the valve while the spring compressor is released

13.3 Remove the bolts that attach the cylinder to the crankcase (arrows)

13.4 Lift the cylinder off and note the location of the dowels (arrows)

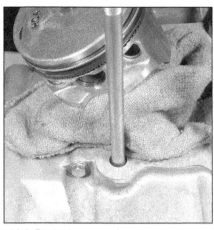

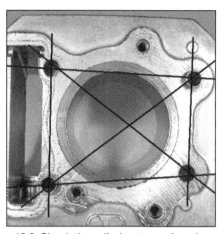

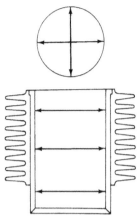

13.5 Pack clean rags into the crankcase opening to keep out debris

13.6 Check the cylinder top surface for warpage in the directions shown

13.8 Measure the cylinder diameter in two directions, at the top, center and bottom of ring travel

5   Locate the dowel pins (they may have come off with the cylinder or still be in the crankcase). Be careful not to let these drop into the engine. Stuff rags around the piston **(see illustration)** and remove the gasket and all traces of old gasket material from the surfaces of the cylinder and the crankcase.

Inspection

Refer to illustrations 13.6 and 13.8

Caution: *Don't attempt to separate the liner from the cylinder.*

6   Check the top surface of the cylinder for warpage, using the same method as for the cylinder head (see Section 12). Measure along the sides and diagonally across the stud holes **(see illustration)**.

7   Check the cylinder walls carefully for scratches and score marks.

8   Using the appropriate precision measuring tools, check the cylinder's diameter at the top, center and bottom of the cylinder bore, parallel to the crankshaft axis **(see illustration)**. Next, measure the cylinder's diameter at the same three locations across the crankshaft axis. Compare the results to this Chapter's Specifications. If the cylinder walls are tapered, out-of-round, worn beyond the specified limits, or badly scuffed or scored, have the cylinder rebored and honed by a dealer service department or an ATV repair shop. If a rebore is done, oversize pistons and rings will be required as well. **Note:** *Honda supplies pistons in two oversizes.*

9   As an alternative, if the precision measuring tools are not available, a dealer service department or other repair shop can make the measurements and offer advice concerning servicing of the cylinder.

10   If it's in reasonably good condition and not worn to the outside of the limits, and if the piston-to-cylinder clearance can be maintained properly, then the cylinder does not have to be rebored; honing is all that is necessary.

11   To perform the honing operation you will need the proper size flexible hone with fine stones as shown in *Maintenance techniques, tools and working facilities* at the front of this book, or a "bottle brush" type hone, plenty of light oil or honing oil, some shop towels and an electric drill motor. Hold the cylinder block in a vise (cushioned with soft jaws or wood blocks) when performing the honing operation. Mount the hone in the drill motor, compress the stones and slip the hone into the cylinder. Lubricate the cylinder thoroughly, turn on the drill and move the hone up and down in the cylinder at a pace which will produce a fine crosshatch pattern on the cylinder wall with the crosshatch lines intersecting at approximately a 60-degree angle. Be sure to use plenty of lubricant and do not take off any more material than is absolutely necessary to produce the desired effect. Do not withdraw the hone from the cylinder while it is running. Instead, shut off the drill and continue moving the hone up and down in the cylinder until it comes to a complete stop, then compress the stones and withdraw the hone. Wipe the oil out of the cylinder and repeat the procedure on the remaining cylinder. Remember, do not remove too much material from the cylinder wall. If you do not have the tools, or do not desire to perform the honing operation, a dealer service department or other repair shop will generally do it for a reasonable fee.

12   Next, the cylinder must be thoroughly washed with warm soapy

13.14 Install the cylinder base gasket and both dowels (arrows) on the crankcase (piston removed for clarity)

13.16 If you're experienced and very careful, the cylinder can be installed over the rings without a ring compressor, but a compressor is recommended

14.3a The IN mark on top of the piston faces the intake (rear) side of the engine

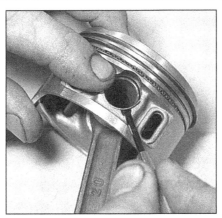

14.3b Wear eye protection and pry the circlip out of its groove with a pointed tool

14.4a Push the piston pin part-way out, then pull it the rest of the way

water to remove all traces of the abrasive grit produced during the honing operation. Be sure to run a brush through the bolt holes and flush them with running water. After rinsing, dry the cylinder thoroughly and apply a coat of light, rust-preventative oil to all machined surfaces.

Installation

Refer to illustrations 13.14 and 13.16

13   Lubricate the cylinder bore with plenty of clean engine oil. Apply a thin film of moly-based grease to the piston skirt.

14   Install the dowel pins, then lower a new cylinder base gasket over them **(see illustration)**.

15   Attach a piston ring compressor to the piston and compress the piston rings. A large hose clamp can be used instead - just make sure it doesn't scratch the piston, and don't tighten it too much.

16   Install the cylinder over the studs and carefully lower it down until the piston crown fits into the cylinder liner **(see illustration)**. While doing this, pull the camshaft chain up, using a hooked tool or a piece of stiff wire. Push down on the cylinder, making sure the piston doesn't get cocked sideways, until the bottom of the cylinder liner slides down past the piston rings. A wood or plastic hammer handle can be used to gently tap the cylinder down, but don't use too much force or the piston will be damaged.

17   Remove the piston ring compressor or hose clamp, being careful not to scratch the piston.

18   The remainder of installation is the reverse of the removal steps.

---

14   Piston - removal, inspection and installation

---

1   The piston is attached to the connecting rod with a piston pin that is a slip fit in the piston and rod.

2   Before removing the piston from the rod, stuff a clean shop towel into the crankcase hole, around the connecting rod. This will prevent the circlips from falling into the crankcase if they are inadvertently dropped.

Removal

Refer to illustrations 14.3a, 14.3b, 14.4a and 14.4b

3   The piston should have an IN mark on its crown that goes toward the intake (rear) side of the engine **(see illustration)**. If this mark is not visible due to carbon buildup, scribe an arrow into the piston crown before removal. Support the piston and pry the circlip out with a pointed tool **(see illustration)**.

4   Push the piston pin out from the opposite end to free the piston from the rod **(see illustration)**. You may have to deburr the area around the groove to enable the pin to slide out (use a triangular file for this procedure). If the pin won't come out, you can fabricate a piston pin removal tool from a long bolt, a nut, a piece of tubing and washers **(see illustration)**.

Inspection

Refer to illustrations 14.6, 14.13, 14.14, 14.15 and 14.16

5   Before the inspection process can be carried out, the pistons must be cleaned and the old piston rings removed.

6   Using a piston ring removal and installation tool, carefully remove the rings from the pistons **(see illustration)**. Do not nick or gouge the pistons in the process.

7   Scrape all traces of carbon from the tops of the pistons. A hand-held wire brush or a piece of fine emery cloth can be used once the majority of the deposits have been scraped away. Do not, under any circumstances, use a wire brush mounted in a drill motor to remove deposits from the pistons; the piston material is soft and will be eroded away by the wire brush.

8   Use a piston ring groove cleaning tool to remove any carbon deposits from the ring grooves. If a tool is not available, a piece broken off the old ring will do the job. Be very careful to remove only the carbon deposits. Do not remove any metal and do not nick or gouge the sides of the ring grooves.

9   Once the deposits have been removed, clean the pistons with solvent and dry them thoroughly. Make sure the oil return holes below the oil ring grooves are clear.

10   If the pistons are not damaged or worn excessively and if the cylinders are not rebored, new pistons will not be necessary. Normal piston wear appears as even, vertical wear on the thrust surfaces of the piston and slight looseness of the top ring in its groove. New piston rings, on the other hand, should always be used when an engine is rebuilt.

11   Carefully inspect each piston for cracks around the skirt, at the pin bosses and at the ring lands.

12   Look for scoring and scuffing on the thrust faces of the skirt, holes in the piston crown and burned areas at the edge of the crown. If the skirt is scored or scuffed, the engine may have been suffering from overheating and/or abnormal combustion, which caused excessively high operating temperatures. The oil pump should be checked thoroughly. A hole in the piston crown, an extreme to be sure, is an indication that abnormal combustion (pre-ignition) was occurring. Burned areas at the edge of the piston crown are usually evidence of spark knock (detonation). If any of the above problems exist, the causes must be corrected or the damage will occur again.

13   Measure the piston ring-to-groove clearance (side clearance) by laying a new piston ring in the ring groove and slipping a feeler gauge in beside it **(see illustration)**. Check the clearance at three or four locations around the groove. Be sure to use the correct ring for each groove; they are different. If the clearance is greater than specified, new pistons will have to be used when the engine is reassembled.

14   Check the piston-to-bore clearance by measuring the bore (see Section 13) and the piston diameter **(see illustration)**. Measure the piston across the skirt on the thrust faces at a 90-degree angle to the piston pin, at the specified distance up from the bottom of the skirt.

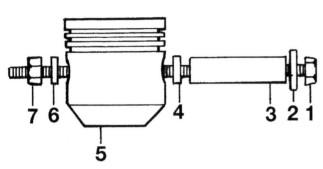

14.4b The piston pin should come out with hand pressure - if it doesn't, this removal tool can be fabricated from readily available parts

1	Bolt	7	Nut (B)
2	Washer	A)	Large enough for
3	Pipe (A)		piston pin to fit inside
4	Padding (A)	B)	Small enough to fit
5	Piston		through piston pin
6	Washer (B)		bore

14.6 Remove the piston rings with a ring removal and installation tool

14.13 Measure the piston ring-to-groove clearance with a feeler gauge

Subtract the piston diameter from the bore diameter to obtain the clearance. If it is greater than specified, the cylinder will have to be rebored and a new oversized piston and rings installed. If the appropriate precision measuring tools are not available, the piston-to-cylinder clearance can be obtained, though not quite as accurately, using feeler gauge stock. Feeler gauge stock comes in 12-inch lengths and various thicknesses and is generally available at auto parts stores. To check the clearance, slip a piece of feeler gauge stock of the same thickness as the specified piston clearance into the cylinder along with thee piston. The cylinder should be upside down and the piston must be positioned exactly as it normally would be. Place the feeler gauge between the piston and cylinder on one of the thrust faces (90-degrees to the piston pin bore). The piston should slip through the cylinder (with the feeler gauge in place) with moderate pressure. If it falls through, or slides through easily, the clearance is excessive and a new piston will be required. If the piston binds at the lower end of the cylinder and is loose toward the top, the cylinder is tapered, and if tight spots are encountered as the piston/feeler gauge is rotated in the cylinder, the cylinder is out-of-round. Be sure to have the cylinder and piston checked by a dealer service department or a repair shop to confirm your findings before purchasing new parts.

15   Apply clean engine oil to the pin, insert it into the piston and check for freeplay by rocking the pin back-and-forth **(see illustration)**. If the

pin is loose, a new piston and possibly a new pin must be installed.
16   Repeat Step 15, this time inserting the piston pin into the connecting rod **(see illustration)**. If the pin is loose, measure the pin diameter and the pin bore in the rod (or have this done by a dealer or repair shop). A worn pin can be replaced separately; if the rod bore is worn, the rod and crankshaft must be replaced as an assembly.
17   Refer to Section 15 and install the rings on the pistons.

14.14 Measure the piston diameter with a micrometer

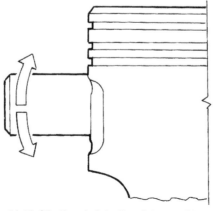

14.15 Slip the pin into the piston and try to wiggle it back-and-forth; if it's loose, replace the piston and pin

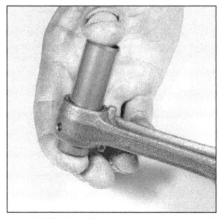

14.16 Slip the piston pin into the rod and try to rock it back-and-forth to check for looseness

14.18 Make sure both piston pin circlips are securely seated in the piston grooves

Installation

Refer to illustration 14.18

18   Install the piston with its IN mark toward the intake side (rear) of the engine. Lubricate the pin and the rod bore with moly-based grease. Install a new circlips in the groove in one side of the piston (don't reuse the old circlips). Push the pin into position from the opposite side and install another new circlip. Compress the circlips only enough for them to fit in the piston. Make sure the clips are properly seated in the grooves **(see illustration)**.

---

15   Piston rings - installation

Refer to illustrations 15.2, 15.4, 15.7a, 15.7b, and 15.7c

1   Before installing the new piston rings, the ring end gaps must be checked.
2   Insert the top (No. 1) ring into the bottom of the first cylinder and square it up with the cylinder walls by pushing it in with the top of the piston. The ring should be about one-half inch above the bottom edge of the cylinder. To measure the end gap, slip a feeler gauge between the ends of the ring **(see illustration)** and compare the measurement to the Specifications.
3   If the gap is larger or smaller than specified, double check to make sure that you have the correct rings before proceeding.
4   If the gap is too small, it must be enlarged or the ring ends may come in contact with each other during engine operation, which can cause serious damage. The end gap can be increased by filing the ring

15.2 Check the piston ring end gap with a feeler gauge at the bottom of the cylinder

15.4 If the end gap is too small, clamp a file in a vise and file the ring ends (from the outside in only) to enlarge the gap slightly

ends very carefully with a fine file **(see illustration)**. When performing this operation, file only from the outside in.
5   Repeat the procedure for the second compression ring (ring gap is not specified for the oil ring rails or spacer).
6   Once the ring end gaps have been checked/corrected, the rings can be installed on the piston.
7   The oil control ring (lowest on the piston) is installed first. It is composed of three separate components. Slip the spacer into the groove, then install the upper side rail **(see illustrations)**. Do not use a

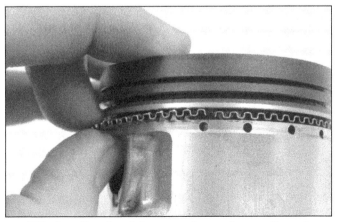

15.7a Installing the oil ring expander - make sure the ends don't overlap

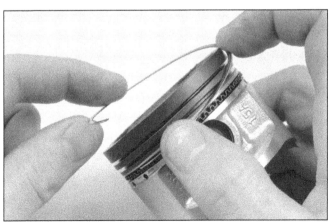

15.7b Installing an oil ring side rail - don't use a ring installation tool to do this

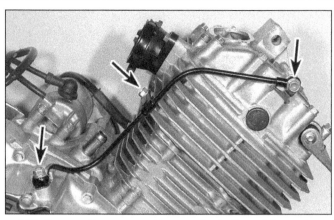

16.2 Remove the upper and lower union bolts and the center retaining bolt (arrows)

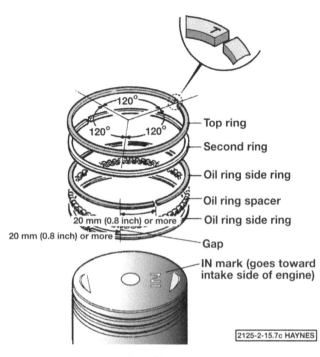

15.7c Ring details

- Top ring
- Second ring
- Oil ring side ring
- Oil ring spacer
- Oil ring side ring
- Gap
- IN mark (goes toward intake side of engine)

20 mm (0.8 inch) or more
20 mm (0.8 inch) or more

2125-2-15.7c HAYNES

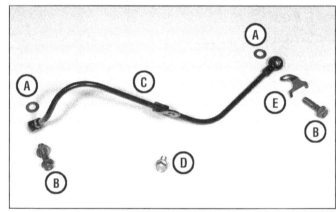

16.3 Oil pipe details

A	Sealing washers	D	Retaining bolt
B	Union bolts	E	Rotation stopper
C	Oil pipe		

piston ring installation tool on the oil ring side rails as they may be damaged. Instead, place one end of the side rail into the groove between the spacer expander and the ring land. Hold it firmly in place and slide a finger around the piston while pushing the rail into the groove (taking care not to cut your fingers on the sharp edges). Next, install the lower side rail in the same manner.

8   After the three oil ring components have been installed, check to make sure that both the upper and lower side rails can be turned smoothly in the ring groove.

9   Install the no. 2 (middle) ring next. It can be readily distinguished from the top ring by its cross-section shape **(see illustration 15.7c)**. Do not mix the top and middle rings.

10   To avoid breaking the ring, use a piston ring installation tool and make sure that the identification mark is facing up **(see illustration 15.7c)**. Fit the ring into the middle groove on the piston. Do not expand the ring any more than is necessary to slide it into place.

11   Finally, install the no. 1 (top) ring in the same manner. Make sure the identifying mark is facing up. Be very careful not to confuse the top and second rings. Besides the different profiles, the top ring is narrower than the second ring.

12   Once the rings have been properly installed, stagger the end gaps, including those of the oil ring side rails **(see illustration 15.7c)**.

16 External oil pipe - removal and installation

Removal

Refer to illustration 16.2

1   Remove the fuel tank (see Chapter 3).

2   To remove the pipe, remove the union bolt at the top and bottom and the retaining bolt in the center of the pipe **(see illustration)**.

Installation

Refer to illustration 16.3

3   Installation is the reverse of the removal steps, with the following additions:

a) *Replace the sealing washers at the upper and lower union bolts whenever the union bolts are loosened* **(see illustration)**.

b) *Position the lower end of the pipe against the cast lug on the crankcase so the pipe won't twist clockwise with the union bolt is tightened.*

c) *Install the metal rotation stopper on the upper end of the pipe, again so the pipe won't twist clockwise when the union bolt is tightened.*

d) *Tighten the union bolts and the retaining bolt to the torques listed in this Chapter's Specifications.*

17 Centrifugal clutch - removal, inspection and installation

Removal

Refer to illustrations 17.8, 17.9 and 17.11

1   Place the shift pedal in the Neutral position.

2   Drain the engine oil (see Chapter 1).

3   Label and disconnect the wires from the reverse, neutral and oil temperature warning switches (see Chapter 8). **Warning:** *Label the reverse and neutral switch wires so there's no chance of mixing them up when they're reconnected. If this happens, the neutral light will come on when the transmission is actually in Reverse. This may cause the vehicle to move backward when you don't expect it.*

4   Remove the reverse stopper lever and cable bracket (see Section 19).

5   Remove the right footpeg (see Chapter 7).

6   Remove the kickstarter pedal (see Section 22).

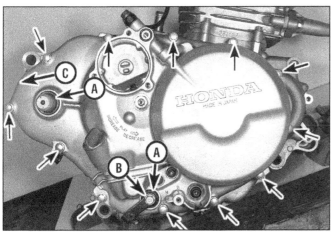

17.8 Seals (A), reverse lock lever (B) and cover bolts; 1998-on models have an extra cover bolt (C)

17.9 Note the locations of the cover dowels (arrows)

17.11 Remove the washer; its OUT SIDE mark (arrow) faces away from the engine

17.13 While holding the drum, it should be possible to turn the weight assembly only in the direction of the arrow; if it turns both ways or neither way, remove it and inspect the one-way clutch

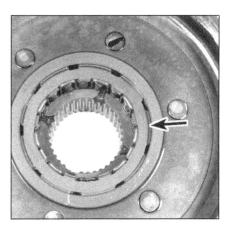

17.15 Lift the one-way clutch (arrow) out of the drum and check it for wear and damage (note that the OUT SIDE mark faces out)

7   If you're working on a 4WD model, remove the skid plate from under the vehicle.

8   Check for signs of oil leakage at the kickstarter seal and reverse lock seal **(see illustration)**. If the seals have been leaking, replace them as described below. Unbolt the right crankcase cover from the engine. Carefully pull the cover off so the kickstarter and reverse lock mechanism aren't pulled out of their holes in the crankcase. If the cover is stuck, tap it gently with a soft face hammer to free it - don't pry the cover off or the gasket surfaces will be damaged.

9   Locate the cover dowels **(see illustration)**. Set them aside for safekeeping.

10   Grind or file away the staked portion of the clutch nut, taking care not to get metal particles in the engine **(see illustration 18.8c)**.

11   Remove the nut, then the washer **(see illustration)**. **Note:** *The nut has left-hand threads. Turn it clockwise to loosen it.*

12   Pull the centrifugal clutch weight assembly and drum off the crankshaft.

Inspection

Refer to illustrations 17.13, 17.15, 17.16, 17.17a, 17.17b, 17.17c, 17.17d, 17.18, 17.25, 17.26, 17.27, 17.31a and 17.31b

13   Hold the drum in one hand and try to rotate the weight assembly with the other hand **(see illustration)**. It should rotate counter-clockwise (anti-clockwise) only. If it rotates both ways or neither way, disassemble the centrifugal clutch for further inspection.

14   Lift the weight assembly out of the drum.

15   Lift the one-way clutch out of the drum **(see illustration)**. Check

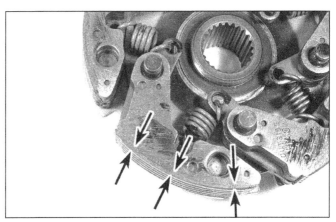

17.16 Measure the thickness of the friction material at the three thick points on each weight (arrows)

the one-way clutch rollers for signs of wear or scoring. The rotors should be unmarked with no signs of wear such as pitting or flat spots. Replace the one-way clutch if it's worn.

16   Measure the thickness of the lining material on the weights at the three thickest points **(see illustration)**. If it's thinner than the minimum listed in this Chapter's Specifications, replace the weights as a set.

17.17a   **Pry the snap-rings loose from the posts**

17.17b   **Lift off the outer plate . . .**

17.17c   **. . . the clutch spring (arrow) . . .**

17.17d   **. . . and the inner plate; the lip on the inner plate (arrow) faces away from the weight assembly**

17.18   **Remove the springs and slip the weights off the drive plate posts; the OUT SIDE mark on each weight (arrow) faces away from the drive plate**

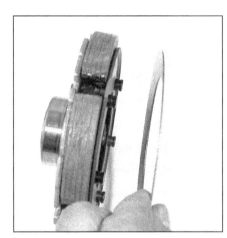

17.25   **Install the clutch spring with its concave side toward the weight assembly**

17   Pry the clips off the drive plate posts **(see illustration)**. Lift off the outer plate, clutch spring and inner plate **(see illustrations)**.

18   Check the springs for breakage and the weights for wear or damage and replace them as necessary. If the weights need to be removed from the drive plate, unhook the springs **(see illustration)** and lift the weights off the posts.

19   Measure the inside diameter of the drum and replace if it it's greater than the limit listed in this Chapter's Specifications.

20   Measure the free length of the clutch weight springs and replace them if they've stretched to longer than the limit listed in this Chapter's Specifications.

21   Check the drive plate posts for wear or damage and replace the drive plate if problems are found.

22   Place the clutch spring on a flat surface (such as a piece of glass) with its concave side down. Measure the height of the spring center from the surface with a vernier caliper. If the spring has been flattened to less than the limit listed in this Chapter's Specifications, replace it.

23   Place the weights in the drive plate with their OUT SIDE marks facing upward **(see illustration 17.18)**. Install the springs, noting how their ends are located.

24   Place the inner plate on the weights with its lip upward **(see illustration 17.17d)**.

25   Place the clutch spring on the inner plate with its concave side toward the plate **(see illustration)**.

26   Place the outer plate on the clutch spring with its locating pins upward **(see illustration)**.

27   Place a pair of C-clamps (G-clamps) or similar clamps on the

17.26   **Install the outer plate with its locating pins (arrows) facing away from the weight assembly**

outer plate and tighten them until the circlip grooves in the posts are exposed, then squeeze the clips onto the posts with pliers **(see illustration)**.

28   Install the one-way clutch in the drum with its OUT SIDE marking facing out **(see illustration 17.15)**. Lubricate the one-way clutch with engine oil.

17.27 Place a pair of clamps on the assembly and compress it so the post grooves are exposed, then squeeze the snap-rings into place with a pair of pliers

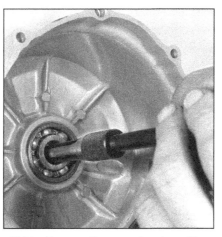

17.31a Remove the cover bearing with a slide hammer and bearing puller attachment . . .

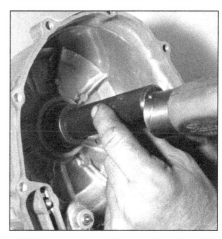

17.31b . . . and install a new bearing with a bearing driver or socket the same diameter as the outer race

17.38 Install a new O-ring (arrow) on the oil pipe fitting

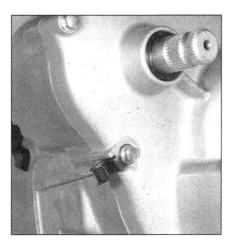

17.39 One cover bolt secures a wiring harness retainer

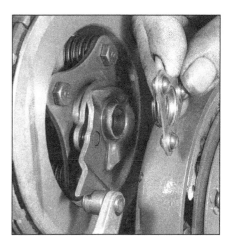

18.2a Remove the ball retainer and spring from the clutch release bearing (if the spring stays in the lifter cam, pull it out) . . .

29   Install the weight assembly in the drum (see illustration 17.13).
30   If the cover seals were leaking, pry them out of their bores (see illustration). Tap in new seals with a seal driver or socket the same diameter as the seal.
31   Rotate the inner race of the ball bearing inside the cover with a finger. If the bearing is rough, loose or noisy, remove it with a blind hole puller, then drive in a new bearing with a bearing driver or socket that bears against the bearing outer race (see illustrations).

Installation

Refer to illustrations 17.38 and 17.39

32   Slip the centrifugal clutch onto the crankshaft. Engage the drum splines with the splines on the crankshaft. Rotate the weight assembly slightly and align the drive plate splines with the primary gear, then push the centrifugal clutch all the way on.
33   Install the washer, making sure its OUT SIDE mark faces away from the engine (see illustration 17.11).
34   Install a new locknut and tighten it to the torque listed in this Chapter's Specifications. **Note:** *The locknut has left-hand threads. Turn it counterclockwise (anti-clockwise) to tighten it.*
35   Stake the lip of the locknut into the slot with a hammer and punch.
36   Make sure the kickstarter, change clutch adjuster and reverse

lock mechanism are in position (see Sections 22, 18 and 19). Make sure the oil strainer screen is in position (see *Engine oil/filter - change* in Chapter 1).
37   Make sure the dowels are in position and install a new gasket (see illustration 17.9).
38   Place a new O-ring on the oil pipe (see illustration).
39   Thread the cover bolts into their holes; note that one bolt secures a wiring harness retainer (see illustration). Tighten the cover bolts in two or three stages, in a criss-cross pattern, to the torque listed in this Chapter's Specifications.
40   The remainder of installation is the reverse of the removal steps.
41   Refill the engine with oil (see Chapter 1).

18   Change clutch - removal, inspection and installation

Release mechanism

Removal

Refer to illustrations 18.2a through 18.2e

1   Remove the right crankcase cover (see Section 17).
2   Remove the release mechanism components from the crankcase and the cover (see illustrations).

18.2b ... pull the clutch lever from the shift pedal shaft; the shaft punch mark aligns with the case arrowhead on installation (arrows) ...

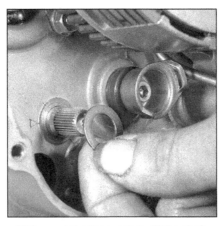

18.2c ... take the washer off the shaft, then remove the lifter cam from the release bearing

18.2d Unscrew the adjuster bolt locknut and remove the washer from the outside of the cover ...

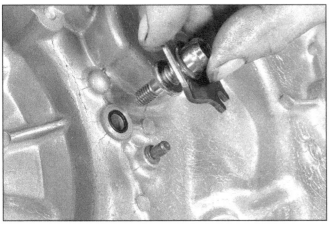

18.2e ... and remove the lifter plate, adjuster bolt and O-ring from the inside of the cover

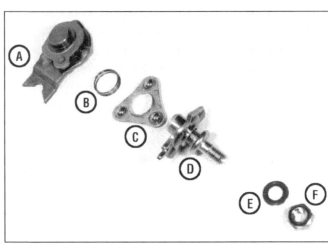

18.3a Release mechanism details

A Lifter cam	D Adjuster bolt
B Spring	E Washer
C Ball retainer	F Adjuster bolt locknut

Inspection

Refer to illustrations 18.3a and 18.3b

3   Check for visible wear or damage at the contact points of the lever and cam and the friction points of the cam, ball retainer and lifter plate **(see illustrations)**. Check the spring for bending or distortion. Replace any parts that show problems. Replace the lifter plate O-ring in the crankcase cover whenever it's removed.

Installation

Refer to illustrations 18.4a and 18.4b

4   Installation is the reverse of the removal steps, with the following additions:

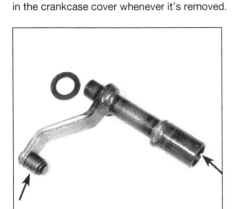

18.3b Check the clutch lever for a worn roller (left arrow) and damaged splines (right arrow)

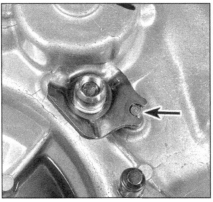

18.4a Make sure the adjusting bolt engages the pin in the crankcase cover (arrow)

18.4b The assembled release mechanism should look like this

18.8a Remove the four bolts and the lifter plate (arrow) . . .

18.8b . . . and remove the pressure plate springs

18.8c Grind or file away the staked portion of the locknut (arrow)

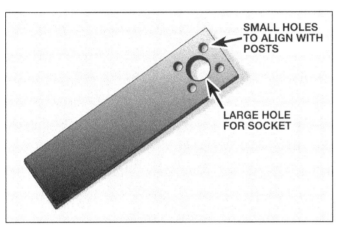

18.8d Clutch holder details

SMALL HOLES
TO ALIGN WITH
POSTS

LARGE HOLE
FOR SOCKET

18.8e Remove the washer; the OUT SIDE mark faces away from the engine on installation

a) *Make sure the lifter plate engages the pin inside the crankcase cover* **(see illustration)**.
b) *Make sure the lever engages the cam* **(see illustration)**.

5   Refill the engine oil and adjust the clutch (see Chapter 1).

Clutch

Removal

Refer to illustrations 18.8a through 18.8k

6   Remove the right crankcase cover and the centrifugal clutch (see

Section 17).

7   Remove the release mechanism as described previously.

8   Refer to the accompanying illustrations to remove the clutch components **(see illustrations)**. Note that if an air wrench is not available, the clutch will have to be prevented from rotating while removing the locknut. The Honda service tool (part no. 07HGB-0010000A) resembles a steering wheel puller and is attached to the spring posts on the pressure plate. It is then held with a breaker bar while the nut is loosened with a socket. A similar tool can be fabricated from metal stock **(see illustration)**.

18.8f Remove the outer friction plate . . .

18.8g . . . then the metal plate, then remove the remaining friction and metal plates . . .

18.8h Remove the pressure plate from the clutch housing

18.8i Remove the thrust washer from the mainshaft . . .

18.8j . . . then remove the clutch housing/primary driven gear . . .

18.8k . . . and the clutch outer guide

Inspection

Refer to illustrations 18.9, 18.11, 18.12, 18.13 and 18.16

9   Check the bolt posts and the friction surface on the pressure plate for damaged threads, scoring or wear **(see illustration)**. Replace the pressure plate if any defects are found.

10   Check the edges of the slots in the clutch housing for indentations made by the friction plate tabs **(see illustration 18.9)**. If the indentations are deep they can prevent clutch release, so the housing should be replaced with a new one. If the indentations can be removed easily with a file, the life of the housing can be prolonged to an extent. Also, check the driven gear teeth for cracks, chips and excessive wear and the springs on the back side for breakage. If the gear is worn or damaged or the springs are broken, the clutch housing must be replaced with a new one. Check the bearing surface in the center of the clutch housing for score marks, scratches and excessive wear.

11   Measure the free length of the clutch springs **(see illustration)** and compare the results to this Chapter's Specifications. If the springs have sagged, or if cracks are noted, replace them with new ones as a set.

12   If the lining material of the friction plates smells burnt or if it is glazed, new parts are required. If the metal clutch plates are scored or discolored, they must be replaced with new ones. Measure the thickness of the friction plates **(see illustration)** and replace with new parts any friction plates that are worn. **Note:** *If any friction plates are worn beyond the minimum thickness, it's a good idea to replace them all.*

13   Lay the metal plates, one at a time, on a perfectly flat surface (such as a piece of plate glass) and check for warpage by trying to slip

18.9 Clutch inspection points

A Pressure plate posts	D Clutch housing bearing
B Pressure plate friction	surface
surface	E Primary driven gear
C Clutch housing slots	

a feeler gauge between the flat surface and the plate **(see illustration)**. The feeler gauge should be the same thickness as the maximum warpage listed in this Chapter's Specifications. Do this at several places around the plate's circumference. If the feeler gauge can be

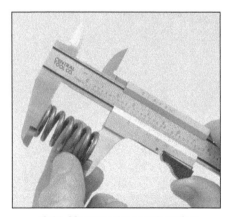

18.11 Measure the clutch spring free length

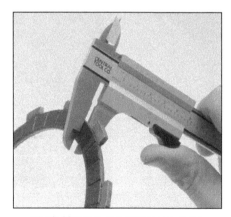

18.12 Measure the thickness of the friction plates

18.13 Check the metal plates for warpage

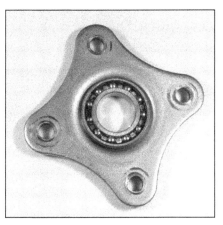

18.16 Check the ball bearing in the center of the lifter plate for roughness, looseness or noise

18.20a Install a friction plate first, then a metal plate, then alternate the remaining friction and metal plates . . .

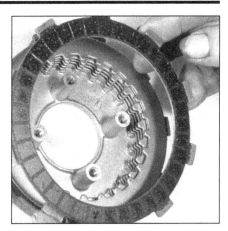

18.20b . . . a friction plate goes on last

18.21a Install the clutch center . . .

18.21b . . . then install the assembly on the mainshaft

18.24 Stake the new locknut

slipped under the plate, it is warped and should be replaced with a new one.

14   Check the tabs on the friction plates for excessive wear and mushroomed edges. They can be cleaned up with a file if the deformation is not severe. Check the friction plates for warpage as described in Step 13.

15   Check the clutch outer guide for score marks, heat discoloration and evidence of excessive wear. Measure its inner and outer diameter and compare them to the values listed in this Chapter's Specifications (a dealer can do this if you don't have precision measuring equipment). Replace the outer guide if it's worn beyond the specified limits. Also measure the end of the transmission mainshaft where the outer guide rides; if it's worn to less than the limit listed in this Chapter's Specifications, replace the mainshaft (see Section 26).

16   Check the clutch lifter plate for wear and damage. Rotate the inner race of the bearing and check for roughness, looseness or excessive noise **(see illustration)**.

17   Check the splines of the clutch center for wear or damage and replace the clutch center if problems are found.

Installation

Refer to illustrations 18.20a, 18.20b, 18.21a, 18.21b and 18.24

18   Lubricate the inner and outer surfaces of the clutch outer guide with moly-based grease and install it on the crankshaft.

19   Install the clutch housing and thrust washer on the mainshaft **(see illustrations 18.8j and 18.8l)**.

20   Coat the friction plates with engine oil. Install a friction plate on

the disc, followed by a metal plate, then alternate the remaining friction and metal plates until they're all installed (there are six friction plates and five metal plates). Friction plates go on first and last, so the friction material contacts the metal surfaces of the clutch center and the pressure plate **(see illustrations)**.

21   Install the clutch center over the posts, then install the assembly in the clutch housing **(see illustrations)**.

22   Install the washer on the mainshaft with its OUT SIDE mark facing away from the engine **(see illustration 18.8e)**.

23   Coat the threads of a new locknut with non-hardening thread locking agent and install it on the mainshaft. Hold the clutch with one of the methods described in Step 8 and tighten the locknut to the torque listed in this Chapter's Specifications.

24   Stake the locknut with a hammer and punch **(see illustration)**.

25   Install the clutch springs and the lifter plate **(see illustrations 18.8b and 18.8a)**. Tighten the bolts to the torque listed in this Chapter's Specifications in two or three stages, in a criss-cross pattern.

26   The remainder of installation is the reverse of the removal steps.

---

19   Reverse lock mechanism - cable replacement, removal, inspection and installation

---

Cable replacement

Refer to illustrations 19.1 and 19.2

1   Loosen the locknut and adjusting nut at the lower end of the cable

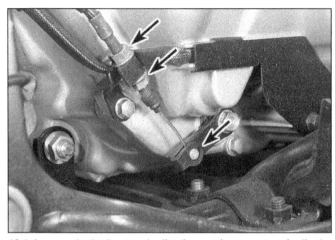

19.1 Loosen the locknut and adjusting nut (upper arrows), slip the cable out of the bracket, then rotate the cable to align it with the lever slot (lower arrow) and slip it out of the lever

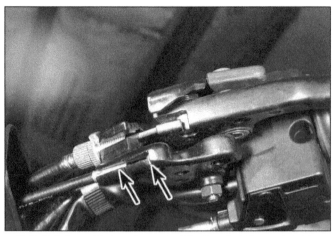

19.2 Pull the cable housing out of the bracket, then rotate the cable out of the horizontal slot (left arrow), align it with the vertical slot (right arrow) and lower it clear

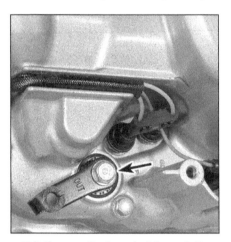

19.5 Remove the lever bolt (arrow); the lever OUT mark faces away from the engine on installation

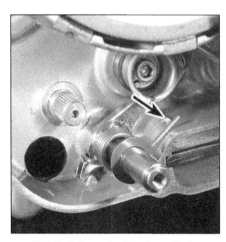

19.7 Pull the end of the lever in the direction of the arrow to disengage it from the lockplate, then pull the shaft out

19.8a Remove the rotor Allen bolt and take the rotor off; the small and large projections (arrows) face in the directions shown on installation

(see illustration). Slip the cable housing out of the bracket, then rotate the cable to align it with the slot in the lever and slip the cable out of the lever.

2   Pull the cable housing out of the handlebar bracket **(see illustration)**. Rotate the cable out of the bracket, then align the cable with the slot in the bracket and slip the cable end down out of the bracket.

3   Note the location of the cable along the frame. Detach it from its retainers and install the new cable in the same location, making sure to secure it with all of the retainers.

4   Reconnect the ends of the cable at the handlebar, lower bracket and lever. Refer to Chapter 1 and adjust the cable.

Removal

Refer to illustrations 19.5, 19.7, 19.8a and 19.8b

5   Remove the bolt and take the lever off the shaft **(see illustration)**.

6   Remove the right crankcase cover (see Section 17).

7   Rotate the lever clockwise to disengage it from the lockplate, then remove the shaft from the crankcase **(see illustration)**.

8   If the switch rotor and lockplate must be removed, remove the centrifugal clutch and change clutch (see Sections 17 and 18). Remove the rotor Allen bolt, then remove the rotor and lockplate **(see illustrations)**. Pull the lockplate pins out of the shift drum if they're worn or damaged.

19.8b Remove the lockplate (arrow); remove the two pins if necessary

Inspection

Refer to illustration 19.9

9   Slide the thrust washer off the shaft **(see illustration)**. Inspect the shaft and lever components and replace any worn or damaged parts. Leave the snap-ring on the shaft unless the shaft, snap-ring or spring needs to be replaced.

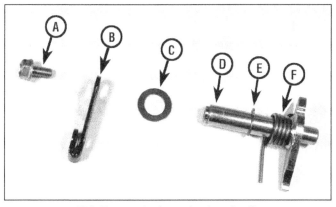

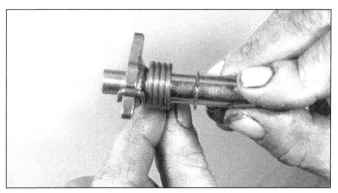

19.10 Hook the end of the spring into the lever notch

19.9 Reverse lock mechanism details

A	Bolt	D	Shaft/internal lever
B	External lever	E	Snap-ring
C	Thrust washer	F	Spring

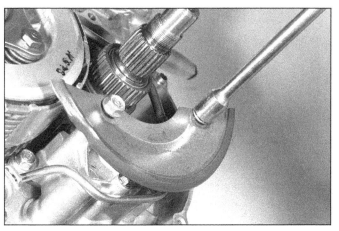

20.1 Remove the separator plate

20.2 Remove the dark-colored bolt (A), the bright-colored bolts (B) and slide the pipe off the fitting (C); use a new O-ring on installation and be sure to put the dark-colored bolt back in its original location

Installation

Refer to illustration 19.10

10   Installation is the reverse of the removal steps, with the following additions:

a) *If the lockplate pins were removed, install them, thicker ends first.*

b) *Apply non-permanent thread locking agent to the threads of the rotor bolt. Align the rotor with the two pins in the lockplate and position its smaller projection downward **(see illustration 19.8a)**. Tighten the rotor bolt to the torque listed in this Chapter's Specifications.*

c) *Lubricate the inner end of the shaft with oil. Hook the spring into the notch in the lever **(see illustration)**, then install the shaft and position the other end of the spring against the crankcase.*

d) *Install the thrust washer on the shaft and slide it against the snapring.*

e) *Install the outer lever with its OUT mark facing away from the engine **(see illustration 19.5)**. Tighten the lever bolt securely, but don't overtighten it.*

---

20 Oil pipe and pump - removal, inspection and installation

Note: *The oil pump can be removed with the engine in the frame.*

Removal

Refer to illustrations 20.1, 20.2, 20.3a and 20.3b

1   Remove the centrifugal clutch and change clutch (see Sec-

20.3a Remove the oil pump mounting bolts (arrows) . . .

tions 17 and 18). Unbolt the separator plate **(see illustration)**.

2   Remove the union bolt and hex bolts that secure the oil pipe **(see illustration)**. Carefully work the oil pipe off the upper fitting and remove it.

3   Remove the mounting bolts, then pull the pump out of the engine **(see illustrations)**.

20.3b . . . and pull the pump off the crankcase; note the locations of the dowels, which may stay in the crankcase or come off with the oil pump, and remove the O-ring (arrows)

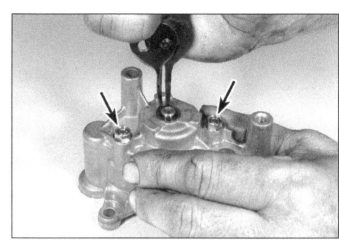

20.4a Remove the snap-ring, washer and bolts (arrows)

20.4b Lift the pump cover off and note the locations of the dowels (arrows)

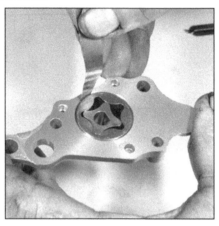

20.7a Measure the clearance between the outer rotor and body . . .

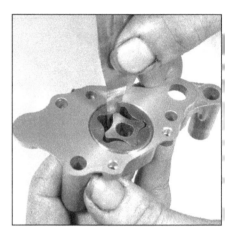

20.7b . . . between the inner and outer rotors . . .

Inspection

Refer to illustrations 20.4a, 20.4b, 20.7a, 20.7b and 20.7c

4   Remove the snap-ring and washer and unbolt the pump cover **(see illustrations)**.

5   Remove the rotors and shaft.

6   Wash all the components in solvent, then dry them off. Check the pump body, the rotors and the cover for scoring and wear. If any damage or uneven or excessive wear is evident, replace the pump. If you are rebuilding the engine, it's a good idea to install a new oil pump.

7   Place the rotors in the pump cover. Measure the clearance between the outer rotor and body, and between the inner and outer rotors, with a feeler gauge **(see illustrations)**. Place a straightedge across the pump body and rotors and measure the gap with a feeler gauge **(see illustration)**. If any of the clearances are beyond the limits listed in this Chapter's Specifications, replace the pump.

8   Reassemble the pump by reversing the disassembly steps, with the following additions:

   a) *Before installing the cover, pack the cavities between the rotors with petroleum jelly - this will ensure the pump develops suction quickly and begins oil circulation as soon as the engine is started.*

   b) *Make sure the cover dowels are in position* **(see illustration 20.4b).**

   c) *If you're working on a 1988 through 1993 model, tighten the cover screws securely.*

   d) *If you're working on a 1994 or later model, tighten the cover bolts to the torque listed in this Chapter's Specifications.*

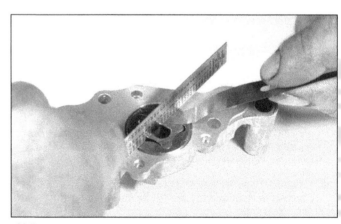

20.7c . . . and between the rotors and a straightedge laid across the pump body

   e) *Install the snap-ring with its chamfered edge toward the oil pump (against the washer).*

Installation

9   Installation is the reverse of removal, with the following additions:

   a) *Make sure the oil pump dowels are in position and install a new O-ring* **(see illustration 20.3b).**

21.2a Slide off the primary drive gear . . .

21.2b . . . and the washer; the friction points (arrows) are measured for wear

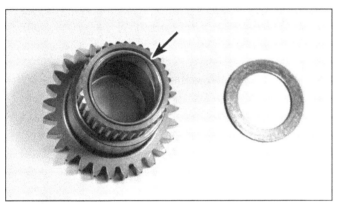

21.4 Measure the diameter of the bushings (arrow); there's one inside each end of the gear

b) *Tighten the oil pump mounting bolts securely, but don't over-tighten them.*

c) *Install the oil pipe bolts in the correct locations* **(see illustra-tion 20.2)**. *Tighten the dark-colored bolt to the torque listed in this Chapter's Specifications. Tighten the other bolts securely but don't overtighten them.*

21 Primary drive gear - removal, inspection and installation

Removal

Refer to illustrations 21.2a and 21.2b

1   Remove the centrifugal clutch, change clutch and oil pump (Sections 17, 18 and 20).

2   Slide the primary drive gear off the crankshaft and remove the washer **(see illustrations)**.

Inspection

Refer to illustration 21.4

3   Check the drive gear for obvious damage such as a worn inner bushing, damaged splines and chipped or broken teeth. Replace it if any of these problems are found.

4   Measure the inside diameter of the bushing at the outer end and inner end of the gear **(see illustration)**. A Honda dealer or machine shop can do this if you don't have precision measuring equipment. If either bushing is worn beyond the diameter listed in this Chapter's Specifications, replace the gear.

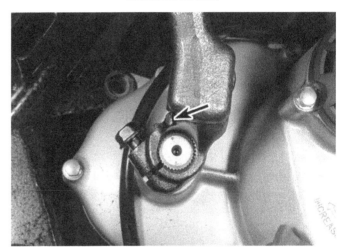

22.2 The punch mark on the end of the spindle aligns with the cast protrusion on the pedal (arrow)

5   Measure the crankshaft at the two points where the bushings ride **(see illustration 21.2b)**. If either point is worn to less than the diameter listed in this Chapter's Specifications, replace the crankshaft (Section 28).

Installation

6   Installation is the reverse of the removal steps.

22 Kickstarter - removal, inspection and installation

Removal

Refer to illustrations 22.2, 22.4a, 22.4b and 22.4c

1   Remove the kickstarter access cover from the right side of the vehicle (see Chapter 7).

2   Look for a punch mark on the end of the kickstarter spindle. The mark should be upward and should align with the cast line on the pedal **(see illustration)**. Loosen the pinch bolt and slide the pedal off the spindle.

3   Remove the right crankcase cover (see Section 17).

4   Slip the pedal back onto the spindle. Rotate it clockwise to free the ratchet from its guide, then unhook the spring and pull the kickstarter assembly out of the crankcase **(see illustrations)**. Locate the inner washer; it may be on the spindle or it may still be in the crankcase **(see illustration)**.

22.4a Disengage the ratchet from its guide (lower arrow) and unhook the spring (upper arrow)

22.4b Pull the kickstarter assembly out of the crankcase

22.4c The inner washer may come out with the kickstarter or remain in the engine

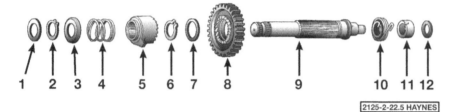

1   2   3   4   5   6   7   8   9   10 11 12

2125-2-22.5 HAYNES

22.5 Kickstarter - exploded view

1	Large washer	5	Ratchet	9	Spindle
2	Small snap-ring	6	Large snap-ring	10	Return spring
3	Spring seat	7	Medium-sized washer	11	Collar
4	Spring	8	Pinion	12	Small washer

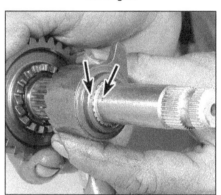

22.8a Align the punch marks on the ratchet and spindle (arrows)

Inspection

Refer to illustrations 22.5, 22.8a, 22.8b, 22.8c and 22.8d

5   Remove the large washer and snap-ring from the outer end of the spindle **(see illustration)**. Remove the spring seat, spring and ratchet. Remove the large snap-ring, washer and pinion. From the inner end of the spindle, remove the small washer, collar and return spring.

6   Check all parts for wear or damage, paying special attention to the teeth on the ratchet and the matching teeth on the pinion. Replace worn or damaged parts.

7   Measure the inside diameter of the pinion and the outside diameter of the shaft where the pinion rides. If either is worn beyond the limit listed in this Chapter's Specifications, replace the worn part.

8   Apply moly-based grease to the splines on the spindle and pinion.

22.8b Place the cupped side of the spring seat against the spring

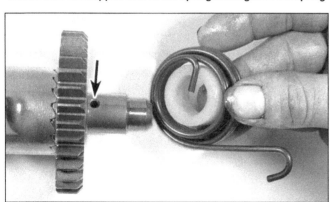

22.8c Fit the spring into the collar notch and place its end in the spindle hole (arrow)

22.8d The assembled kickstarter should look like this

23.1 Check for alignment marks on the shift shaft and pedal (arrows); make your own marks if they aren't visible

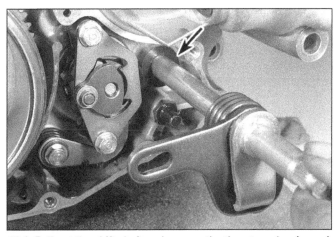

23.6 Pull out the shift shaft and remove the thrust washer (arrow)

Reassemble the kickstarter **(see illustration 22.5)**, noting the following:

a) *Align the punch marks on the ratchet and spindle* **(see illustration)**.
b) *Place the cupped side of the spring seat over the spring* **(see illustration)**.
c) *Align the end of the return spring with the collar groove and place the end of the return spring in the spindle hole* **(see illustrations)**.

Installation

9   Slip the pedal onto the kickstarter spindle. Place the kickstarter in position in the crankcase. Hook the spring onto its post, then rotate the kickstarter clockwise with the pedal to engage the ratchet with the ratchet guide and push the kickstarter the rest of the way in **(see illustration 22.4a)**.
10   The punch mark on the outer end of the spindle should now be straight up **(see illustration 22.2)**. If it isn't the kickstarter is assembled or installed incorrectly; find the problem before proceeding.
11   The remainder of installation is the reverse of the removal steps.

---

23 External shift mechanism - removal, inspection and installation

Shift pedal

Removal

Refer to illustration 23.1

1   Look for alignment marks on the end of the shift pedal and shift shaft **(see illustration)**. If they aren't visible, make your own marks with a sharp punch.
2   Remove the shift pedal pinch bolt and slide the pedal off the shaft.

Inspection

3   Check the shift pedal for wear or damage such as bending. Check the splines on the shift pedal and shaft for stripping or step wear. Replace the pedal or shaft if these problems are found.

Installation

4   Install the shift pedal. Line up the punch marks and tighten the pinch bolt to the torque listed in this Chapter's Specifications.

External shift linkage

Removal

Refer to illustrations 23.6, 23.7, 23.8a, 23.8b, 23.9, 23.10 and 23.11

5   Remove the right crankcase cover (see Section 17) and the

23.7 Shift mechanism details

A	Guide plate bolts	D	Drum shifter
B	Guide plate	E	Return spring post
C	Collar		

23.8a Remove the collars (arrows) . . .

alternator cover (see Chapter 8).
6   Pull the shift shaft out of the crankcase and take its washer off **(see illustration)**.
7   Unbolt the guide plate and lift it off, together with the collar and drum shifter **(see illustration)**.
8   Remove the collars, dowels and bearing stopper plates **(see illustrations)**.

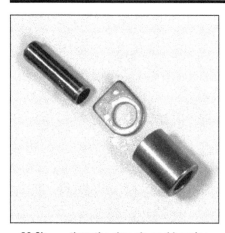

23.8b ... then the dowels and bearing stopper plates

23.9 Unscrew the drum center bolt

23.10 Unbolt the stopper arm and remove it together with the spring

23.11 Lift off the drum center; align the dowel and notch on installation (arrows)

23.12 The ratchet pawls fit in the drum center like this, with the rounded ends of the pawls facing each other

23.13a Pry apart the return spring ...

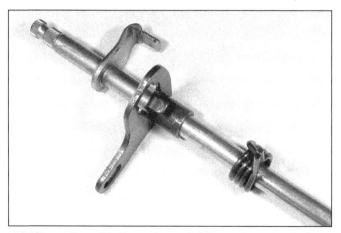

23.13b ... then slide the spring and shift arm down the shaft so they can be inspected

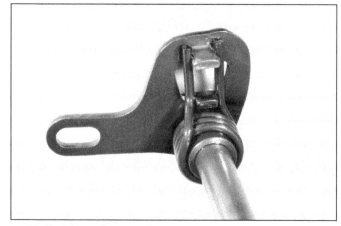

23.13c The spring ends fit over the shift arm and shift shaft like this

9   Remove the bolt from the shift drum center **(see illustration)**.

10   Unbolt the stopper arm, then remove the arm and its spring **(see illustrations)**.

11   Remove the shift drum center and note the location of its dowel **(see illustration)**.

Inspection

Refer to illustrations 23.12, 23.13a, 23.13b and 23.13c

12   Remove the drum shifter from the guide plate **(see illustration)**.

Remove the ratchet pawls, plungers and springs from the drum shifter **(see illustration)**. Inspect the parts and replace any that are worn or damaged.

13   Check the shift shaft for bends and damage to the splines. If the shaft is bent, you can attempt to straighten it, but if the splines are damaged it will have to be replaced. Pry the return spring apart, then slide it and the shift arm down the shaft **(see illustrations)**. Check the condition of the shift arm and the pawl spring. Replace them if they're worn, cracked or distorted. If the arm, spring and shaft are good, reassemble them **(see illustration)**.

23.17 The ends of the spring (lower arrows) rest against the crankcase and hook over the stopper arm; the roller end (upper arrows) engages a notch in the drum center

23.18a Install the dowels and position the bearing stoppers over them . . .

23.18b . . . and install the collars over the dowels

23.19a Place the guide plate over the dowels and against the collars . . .

23.19b . . . then install the guide plate bolts . . .

14   Make sure the return spring post isn't loose **(see illustration 23.7)**. If it is, unscrew it, apply a non-hardening locking compound to the threads, then reinstall it and tighten it securely.

15   Inspect, and replace if necessary, the seals and the shift pedal bearing in the right crankcase cover and alternator cover (see Section 17 and Chapter 8).

Installation

Refer to illustrations 23.17, 23.18a, 23.18b, 23.19a, 23.19b, 23.20 and 23.21

16   Position the spring on the stopper arm, then install the stopper arm on the engine and tighten its bolt securely.

17   Pull down the stopper arm and install the drum center on the shift drum, making sure its dowel is located in the drum center notch **(see illustration 23.11)**. Apply non-permanent thread locking agent to the threads of drum center bolt, then install it and tighten it securely. Make sure the stopper arm spring is correctly installed and that the roller end of the stopper arm engages a notch in the drum center **(see illustration)**.

18   Install the dowels and bearing stopper plates in the crankcase, then install the collars over the dowels **(see illustrations)**.

19   Place the assembled guide plate and drum shifter over the dowels, then install the guide plate bolts **(see illustrations)**.

20   Place the shift collar on the drum shifter **(see illustration)**.

23.20 . . . and place the collar on the drum shifter

23.21 The assembled shift linkage should look like this

24.11   Temporarily reinstall the guide plate with its bearing stopper plates, dowels and collars to keep the bearing from falling out of the case

21   Place the thrust washer on the shift shaft and slide it into the engine **(see illustration 23.6)**. Slide the shaft all the way in, making sure the shift arm fits over the shift collar **(see illustration)**.
22   The remainder of installation is the reverse of the removal steps.
23   Check the engine oil level and add some, if necessary (see Chapter 1).

---

24   Crankcase - disassembly and reassembly

1   To examine and repair or replace the crankshaft, connecting rod, bearings and transmission components, the crankcase must be split into two parts.

Disassembly

Refer to illustrations 24.11, 24.12a, 24.12b, 24.12c, 24.13 and 24.14

2   Remove the engine from the vehicle (see Section 5).
3   Remove the carburetor (see Chapter 3).
4   Remove the alternator rotor and the starter motor (see Chapter 8).
5   Remove the centrifugal clutch (see Section 17) and the change clutch (see Section 18).
6   Remove the external shift mechanism (see Section 23).
7   Remove the reverse lock mechanism (see Section 19).
8   Remove the cylinder head cover, cam chain tensioner, cam chain, cylinder head, cylinder and piston (see Sections 7, 8, 9, 10, 13 and 14).
9   Remove the oil pipe and pump (see Section 20).
10   Check carefully to make sure there aren't any remaining components that attach the upper and lower halves of the crankcase together.
11   Temporarily install the guide plate for the external shift linkage, along with its bearing stopper plates, dowels and collars **(see illustration)**. Refer to Section 23 for details if necessary.
12   Loosen the left crankcase bolts in two or three stages, in a criss-cross pattern **(see illustrations)**. Remove the bolts and label them; they are different lengths. Remove the single bolt from the right side of the crankcase **(see illustration)**.
13   Carefully pry the crankcase apart and lift the right half off the left half **(see illustration)**. **Caution:** *Only pry on casting protrusions - don't pry against the mating surfaces or leaks will develop.*
14   Remove the two crankcase dowels **(see illustration)**.
15   Refer to Sections 25 through 28 for information on the internal components of the crankcase.

Reassembly

16   Remove all traces of old gasket and sealant from the crankcase mating surfaces with a sharpening stone or similar tool. Be careful not

24.12a Loosen the left side case bolts (arrows) in two or three stages, in a criss-cross pattern (4WD crankcase shown); bolt A is hidden under the case and bolt B is hidden behind the output gear

to let any fall into the case as this is done and be careful not to damage the mating surfaces.
17   Check to make sure the two dowel pins are in place in their holes in the mating surface of the left crankcase half **(see illustration 24.14)**.
18   Pour some engine oil over the transmission gears, the crankshaft bearing surface and the shift drum. Don't get any oil on the crankcase mating surface.
19   Install a new gasket on the crankcase mating surface.
20   Carefully place the right crankcase half onto the left crankcase half. While doing this, make sure the transmission shafts, shift drum, crankshaft and balancer fit into their ball bearings in the right crankcase half.
21   Install the left crankcase half bolts in the correct holes and tighten them so they are just snug. Then tighten them in two or three stages, in a criss-cross pattern, to the torque listed in this Chapter's Specifications (don't forget to install the hose retainers).
22   Tighten the single bolt in the right half of the crankcase to the torque listed in this Chapter's Specifications.
23   Turn the transmission mainshaft to make sure it turns freely. Also make sure the crankshaft turns freely.
24   The remainder of installation is the reverse of removal.

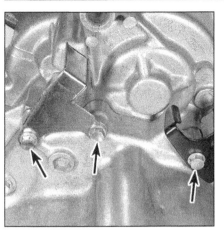

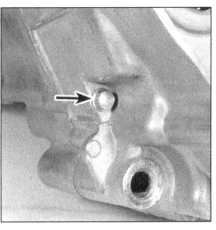

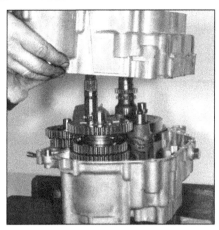

24.12b These three bolts at the upper rear secure hose retainers

24.12c Remove the single bolt from the right side (arrow)

24.13 Lift the right case half off the left half

24.14 Note the locations of the case dowels (arrows)

25.1 Remove the kickstarter ratchet guide (left arrow), the bearing retainer (center arrow) and the cam chain guide cup (right arrow)

25 Crankcase components - inspection and servicing

Refer to illustrations 25.1, 25.3a, 25.3b and 25.3c

1   Separate the crankcase and remove the following:

a) *Transmission shafts and gears*
b) *Output gear*
c) *Crankshaft and main bearings*
d) *Shift drum and forks*
e) *Breather hoses* **(see illustration)**
f) *Kickstarter ratchet guide, mainshaft bearing retainer and cam chain guide cup* **(see illustration)**.

2   Clean the crankcase halves thoroughly with new solvent and dry them with compressed air. All oil passages should be blown out with compressed air and all traces of old gasket sealant should be removed from the mating surfaces. **Caution:** *Be very careful not to nick or gouge the crankcase mating surfaces or leaks will result. Check both crankcase sections very carefully for cracks and other damage.*

3   Check the bearings in the case halves **(see illustration 24.14 and the accompanying illustrations)**. If they don't turn smoothly, replace them. The mainshaft needle roller bearing and balancer ball bearing aren't accessible from the outside, so a blind hole puller will be needed for removal **(see illustration)**. Drive the remaining bearings out with a bearing driver or a socket having an outside diameter slightly smaller than that of the bearing outer race. Before installing the bearings, allow

25.3a Right side case bearings (arrows)

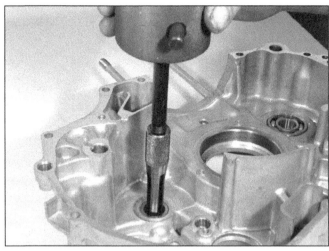

25.3b Bearings that aren't accessible from both sides can be removed with a slide hammer and puller attachment . . .

25.3c . . . and driven in with a bearing driver or socket (ball bearing) or shouldered drift (needle roller bearing); bearings accessible from both sides can be removed and installed with a bearing driver or socket

26.2a Lift up the shift fork shaft (arrow) . . .

26.2b . . . and remove the right shift fork; note that the letter R faces up (shift drum removed for clarity) . . .

26.2c . . . lift the shift drum out of its bearing . . .

them to sit in the freezer overnight, and about fifteen-minutes before installation, place the case half in an oven, set to about 200-degrees F, and allow it to heat up. The bearings are an interference fit, and this will ease installation. **Warning:** *Before heating the case, wash it thoroughly with soap and water so no explosive fumes are present. Also, don't use a flame to heat the case. Install the ball bearings with a socket or bearing driver that contacts the bearing outer race. Install the needle roller bearing with a shouldered drift that fits inside the bearing to keep it from collapsing while the shoulder applies pressure to the outer race.*

4   If any damage is found that can't be repaired, replace the crankcase halves as a set.

5   Assemble the case halves (see Section 24) and check to make sure the crankshaft and the transmission shafts turn freely.

---

26   Transmission shafts and shift drum - removal, inspection and installation

---

Note: *When disassembling the transmission shafts, place the parts on a long rod or thread a wire through them to keep them in order and facing the proper direction.*

Removal

Refer to illustrations 26.2a through 26.2s

1   Remove the engine, then separate the case halves (see Sections 5 and 24).

2   The transmission components and shift drum remain in the left case half when the case is separated. To remove the components, refer to the accompanying photo sequence **(see illustrations)**.

Inspection

Refer to illustrations 26.5 and 26.7

3   Wash all of the components in clean solvent and dry them off. Rotate the countershaft, feeling for tightness, rough spots, excessive looseness and listening for noises in the bearing or output gear. If any of these conditions are found, remove the output gear (see Section 27) and take it to a Honda dealer or other repair shop and have them disassemble it for further inspection.

4   Inspect the shift fork grooves in the countershaft first/reverse shifter, mainshaft third gear and countershaft fourth gear between third and fourth gears. If a groove is worn or scored, replace the affected part and inspect its corresponding shift fork.

26.2d . . . disengaging it from the center fork pin (upper arrow) and left fork pin (lower arrow) . . .

26.2e . . . remove the washer and countershaft first gear (arrows) . . .

26.2f . . . the first gear bushing and the reverse idler gear washer (arrow) . . .

26.2g . . . lift the reverse idler gear (arrow) off the reverse idler shaft . . .

26.2h . . . remove both of the shaft bushings (upper arrows), the washer (lower arrow) and the shaft from the case . . .

26.2i Remove the countershaft first-reverse shifter (arrow) . . .

26.2j . . . the spline collar (arrow) . . .

26.2k ... countershaft reverse gear (arrow) ...

26.2l ... the bushing and countershaft second gear (arrows) ...

26.2m ... the center shift fork (arrow); its letter C faces upward on installation ...

26.2n ... the spline collar (upper arrow) and countershaft fourth gear (lower arrow) ...

26.2o ... the mainshaft, together with fourth gear (arrow), its bushing, spline washer and snap-ring ...

26.2p ... the countershaft bushing and collar (right arrows), third gear and the left shift fork (left arrows); the letter L on the shift fork faces upward on installation ...

26.2q ... the washer and mainshaft fifth gear (arrows) ...

26.2r ... the fifth gear bushing and bearing washer ...

26.2s . . . and countershaft fifth gear (arrow)

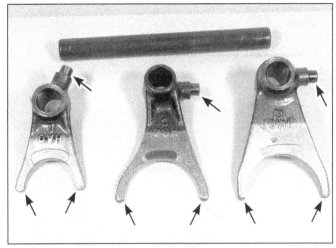

26.5 The fork ears and pins (arrows) are common wear points

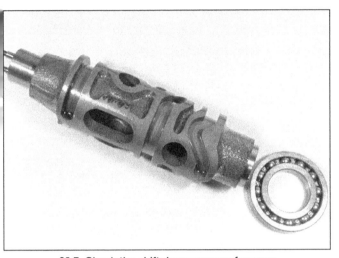

26.7 Check the shift drum grooves for wear

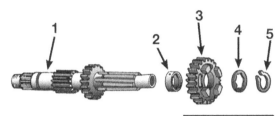

26.12a Install the bushing, fourth gear, spline washer and snap-ring on the mainshaft

1	Mainshaft	4	Spline washer
2	Bushing	5	Snap-ring
3	Fourth gear		

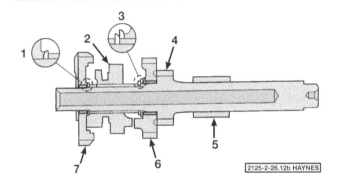

26.12b Mainshaft details

1	Washer	4	Second gear (integral
2	Third gear		with shaft)
3	Snap-ring and spline	5	First gear
	washer	6	Fourth gear
		7	Fifth gear

5   Check the shift forks for distortion and wear, especially at the fork ears **(see illustration)**. Measure the thickness of the fork ears and compare your findings with this Chapter's Specifications. If they are discolored or severely worn they are probably bent. Inspect the guide pins for excessive wear and distortion and replace any defective parts with new ones.

6   Measure the inside diameter of the forks and the outside diameter of the fork shaft and compare to the values listed in this Chapter's Specifications. Replace any parts that are worn beyond the limits. Check the shift fork shaft for evidence of wear, galling and other damage. Make sure the shift forks move smoothly on the shafts. If the shafts are worn or bent, replace them with new ones.

7   Check the edges of the grooves in the drum for signs of excessive wear **(see illustration)**.

8   Hold the inner race of the shift drum bearing with fingers and spin the outer race. Replace the bearing if it's rough, loose or noisy.

9   Check the gear teeth for cracking and other obvious damage. Check the bushing and surface in the inner diameter of the freewheeling gears for scoring or heat discoloration. Replace damaged parts.

10   Inspect the engagement dogs and dog holes on gears so equipped for excessive wear or rounding off. Replace the paired gears as a set if necessary.

11   Check the mainshaft needle bearing in the crankcase for wear or heat discoloration and replace them if necessary (see Section 25).

Installation

Refer to illustrations 26.12a through 26.12e

12   Installation is the basically the reverse of the removal procedure, but take note of the following points:

a) *Use a new snap-ring to secure the spline washer, fourth gear and bushing to the mainshaft* **(see illustration)**. *Install the spline washer and snap-ring, and the fifth gear washer, with their sharp edges facing in the correct directions* **(see illustration)**.

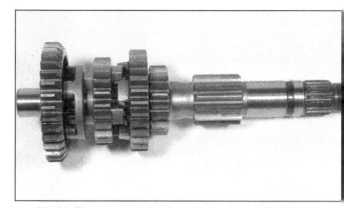

26.12d The assembled mainshaft should look like this . . .

26.12c Place the reverse idler shaft pin (arrow) in the crankcase notch

b) *Place the reverse idler shaft pin in the crankcase notch* **(see illustration)**.
c) *Lubricate the components with engine oil before assembling them.*
d) *After assembly, check the gears to make sure they're installed correctly* **(see illustrations)**.

27 Output gear and countershaft - removal, inspection and installation

Removal

Refer to illustrations 27.2a, 27.2b and 27.3

1   Remove the engine and remove the transmission components from the crankcase (Sections 5 and 26).
2   Check the output gear for leakage at its seals **(see illustrations)**. Rotate the countershaft and check for loose, rough or noisy movement. If any of these conditions exist, remove the output gear and take it to Honda dealer for disassembly and repair.
3   To remove the output gear, unscrew its mounting bolts and pull it off the crankcase **(see illustration 27.2 and the accompanying illustration)**.

Installation

Refer to illustration 27.4

4   Installation is the reverse of removal. Use a new O-ring, lubricated with engine oil **(see illustration)**. Tighten the bolts to the torque listed in this Chapter's Specifications.

26.12e . . . and should mesh with the countershaft like this when they're installed

28 Crankshaft and balancer - removal, inspection and installation

Note: *The procedures in this section require a press and some special tools. If you don't have the necessary equipment or suitable substitutes, have the crankshaft removed and installed by a Honda dealer.*

27.2a Check for leakage at the rear seal . . .

27.2b . . . and at the front seal; to detach the output gear, remove its mounting bolts (arrows) . . .

27.3 . . . and pull it off the crankcase; note the location of the dowel

27.4 Use a new O-ring on installation (arrow); lubricate it with engine oil

28.2a Press the crankshaft out of the left crankcase half . . .

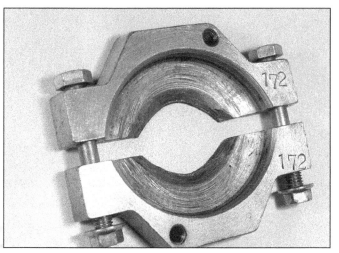

28.2b . . . if the bearing stays on the crankshaft, remove it with the press and a bearing splitter

28.3 Check the connecting rod side clearance with a feeler gauge

Removal

Refer to illustrations 28.2a and 28.2b

1   Remove the engine, separate the crankcase halves and remove the transmission (Sections 5, 24 and 26). The output gear need not be removed.

2   Place the left crankcase half in a press and press out the crankshaft, removing the balancer at the same time **(see illustration)**. The ball bearing may remain in the crankcase or come out with the crankshaft. If it stays on the crankshaft, remove it with the press and a bearing splitter **(see illustration)**. Discard the bearing, no matter what its apparent condition, and use a new one on installation.

Inspection

Refer to illustrations 28.3, 28.4, 28.6 and 28.7

3   Measure the side clearance between connecting rod and crankshaft with a feeler gauge **(see illustration)**. If it's more than the limit listed in this Chapter's Specifications, replace the crankshaft and connecting rod as an assembly.

4   Set up the crankshaft in V-blocks with a dial indicator contacting the big end of the connecting rod **(see illustration)**. Move the connecting rod up-and-down against the indicator pointer and compare the reading to the value listed in this Chapter's Specifications. If it's beyond the limit, replace the crankshaft and connecting rod as an assembly.

5   Check the crankshaft gear, sprockets and bearing journals for

28.4 Check the connecting rod radial clearance with a dial indicator

visible wear or damage, such as chipped teeth or scoring. If any of these conditions are found, replace the crankshaft and connecting rod as an assembly.

6   Set the crankshaft in a pair of V-blocks, with a dial indicator

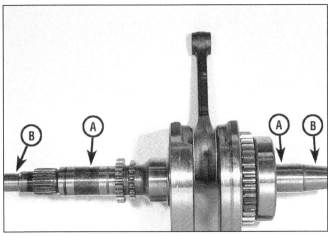

28.6 Place a V-block on each side of the crankshaft (A) and measure runout at the ends (B)

28.7 Check the balancer gear and bearing journals for wear or damage

28.9a Place the crankshaft and balancer in the right crankcase half . . .

28.9b . . . making sure to align the timing marks (arrow)

contacting each end **(see illustration)**. Rotate the crankshaft and note the runout. If the runout at either end is beyond the limit listed in this Chapter's Specifications, replace the crankshaft and connecting rod as an assembly.

7   Check the balancer gear and bearing journals for visible wear or damage and replace the balancer if any problems are found **(see illustration)**.

Installation

Refer to illustrations 28.9a, 28.9b, 28.10, 28.11, 28.12 and 28.14

8   Install the balancer bearing and crankshaft bearing in the left crankcase half (see Section 25).

9   Install the crankshaft and balancer in the right crankcase half with their timing marks aligned **(see illustrations)**. **Note:** *It's important to align the timing marks exactly throughout this procedure. Severe engine vibration will occur if the crankshaft and balancer are out of time.*

10   Thread a puller adapter (Honda part no. 07931-KF00200 in the US or 07965-VM00300 in other markets) into the end of the crankshaft **(see illustration)**.

11   Temporarily install the left crankcase half on the right crankcase **(see illustration)**.

12   Install a crankshaft puller (Honda tools 9765-VM00100 and 09731-ME4000A) over the crankshaft **(see illustration)**.

13   Hold the puller shaft with one wrench and turn the nut with

another wrench to pull the crankshaft into the center race of the ball bearing.

14   Remove the special tools. Lift the right crankcase half off and check to make sure the timing marks on the other side of the crankshaft and balancer are aligned **(see illustration)**.

15   The remainder of installation is the reverse of the removal steps.

29 Initial start-up after overhaul

1   Make sure the engine oil level is correct, then remove the spark plug from the engine. Place the engine kill switch in the Off position and unplug the primary (low tension) wires from the coil.

2   Turn on the key switch and crank the engine over with the starter several times to build up oil pressure. Reinstall the spark plug, connect the wires and turn the switch to On.

3   Make sure there is fuel in the tank, then operate the choke.

4   Start the engine and allow it to run at a moderately fast idle until it reaches operating temperature. **Caution:** *If the oil temperature light doesn't go off, or it comes on while the engine is running, stop the engine immediately.*

5   Check carefully for oil leaks and make sure the transmission and controls, especially the brakes, function properly before road testing the machine. Refer to Section 30 for the recommended break-in procedure.

28.10 Thread the special tool adapter into the end of the crankshaft . . .

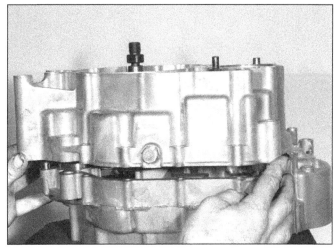

28.11 . . . install the left crankcase half without bolts . . .

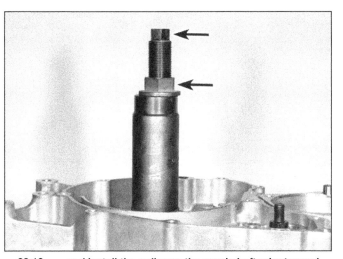

28.12 . . . and install the puller on the crankshaft adapter and crankcase half; hold the shaft (upper arrow) with a wrench and turn the nut to pull the crankshaft into the bearing

28.14 Remove the right crankcase half and make sure the timing marks are aligned (arrows); if they aren't, remove the crankshaft and balancer and reinstall them correctly

6   Upon completion of the road test, and after the engine has cooled down completely, recheck the valve clearances (see Chapter 1).

30 Recommended break-in procedure

1   Any rebuilt engine needs time to break-in, even if parts have been installed in their original locations. For this reason, treat the machine gently for the first few miles to make sure oil has circulated throughout the engine and any new parts installed have started to seat.
2   Even greater care is necessary if the cylinder has been rebored or

a new crankshaft has been installed. In the case of a rebore, the engine will have to be broken in as if the machine were new. This means greater use of the transmission and a restraining hand on the throttle for the first few operating days. There's no point in keeping to any set speed limit - the main idea is to vary the engine speed, keep from lugging (laboring) the engine and to avoid full-throttle operation. These recommendations can be lessened to an extent when only a new crankshaft is installed. Experience is the best guide, since it's easy to tell when an engine is running freely.
3   If a lubrication failure is suspected, stop the engine immediately and try to find the cause. If an engine is run without oil, even for a short period of time, irreparable damage will occur.

Notes

Chapter 3
Fuel and exhaust systems

Contents

	Section
Air cleaner housing - removal and installation	9
Air filter element - servicing	See Chapter 1
Carburetor - removal and installation	6
Carburetor overhaul - general information	5
Carburetors - disassembly, cleaning and inspection	7
Carburetors - reassembly and float height check	8
Choke cable - removal and installation	11
Exhaust system - check	See Chapter 1

	Section
Exhaust system - removal and installation	12
Fuel tank - cleaning and repair	3
Fuel tank - removal and installation	2
General information	1
Idle fuel/air mixture adjustment	4
Spark arrester cleaning	See Chapter 1
Throttle cable - removal, installation and adjustment	10

Specifications

General

Fuel type .... Unleaded or low lead gasoline (petrol) subject to local regulations; minimum octane 91 RON (87 pump octane)

Carburetor

Identification mark

1988 through 1990	VE90A
1991	VE90C
1992	VE90D
1993	VE90E
1994	VE91A
1995 through 1997	VE91C
1998 and later	
Except California	VE91C
California	VE91D

Jet sizes and settings

Standard main jet

1988 through 1990	120
1991 and later	125

High altitude main jet

1988 through 1990	115
1991 and later	120
Jet needle clip position	Third groove from top

Slow jet

1988 through 1990	42
1991 and later	40

Pilot screw setting (turns out from lightly seated position)

Sea level to 5000 feet (1500 meters)

1988 through 1991	1-3/4
1992	1-5/8
1993 through 1997	2-1/4
1998 and later	
Except California	2-1/4
California	2-1/2

Pilot screw setting (continued)
    High altitude (3000 to 8000 feet/1000 to 2500 meters)
    (turns in from standard setting)
        1988 through 1990.. 1/2
        1991 .... 3/4
        1992 .... 1/2
        1993 and later 3/4
Choke (starter) jet
        1988 through 1990.. 85
        1991 and 1992 80
        1993 and later 85
Float level.. 18.5 mm (47/64-inch)

Torque settings

Carburetor cover screw .... 3.5 Nm (30 in-lbs)
Starting enrichment valve nut (1993 and later)...................................... 2.8 Nm (22 in-lbs)
Carburetor band screws... 4 Nm (35 in-lbs)
Muffler mounting bolts.. 55 Nm (40 ft-lbs)
Exhaust pipe heat shield bolts
        1988 through 1992 .... 10 Nm (84 in-lbs)
        1993 and later .... 18 Nm (156 in-lbs)

1   General information

    The fuel system consists of the fuel tank, fuel tap, filter screen, carburetor and connecting lines, hoses and control cables.

    The carburetor used on these vehicles is a constant vacuum unit with a butterfly-type throttle valve. For cold starting, an enrichment circuit is actuated by a cable and the choke lever mounted on the left handlebar.

    The exhaust system consists of a pipe and muffler/silencer with a spark arrester function.

    Many of the fuel system service procedures are considered routine maintenance items and for that reason are included in Chapter 1.

2   Fuel tank - removal and installation

Warning: *Gasoline (petrol) is extremely flammable, so take extra precautions when you work on any part of the fuel system. Don't smoke or allow open flames or bare light bulbs near the work area, and don't work in a garage where a natural gas-type appliance (such as a water heater or clothes dryer) with a pilot light is present. Since gasoline is carcinogenic, wear latex gloves when there's a possibility of being exposed to fuel, and, if you spill any fuel on your skin, rinse it off immediately with soap and water. Mop up any spills immediately and do not store fuel-soaked rags where they could ignite. When you perform any kind of work on the fuel system, wear safety glasses and have a Class B type fire extinguisher on hand.*

1   The fuel tank is secured to a bracket by a bolt and washer at the rear. Two rubber insulators on the bracket support the rear end of the tank. At the front, the tank is supported by cups that fit over rubber mounts.

Removal

Refer to illustrations 2.3, 2.4 and 2.6

2   Remove the seat (see Chapter 7).

3   Remove the fuel tank mounting bolt **(see illustration).**

4   Disconnect the fuel line from the fuel tap **(see illustration)**.

5   Pull the fuel tank backward off the front mounting dampers and lift it off the vehicle together with the fuel tap.

6   If necessary for other procedures, remove the fuel tank bracket **(see illustration)**.

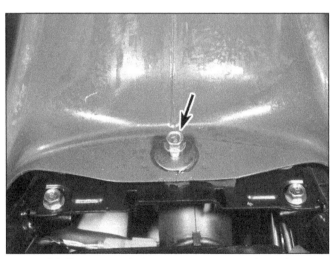

2.3 Remove the fuel tank bolt and washer (arrow)

2.4 Disconnect the fuel line from the tap

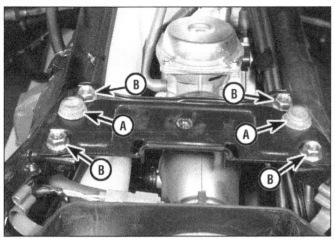

2.6 Inspect the rubber dampers (A) and replace them if they're deteriorated; remove the bolts (B) if it's necessary to remove the fuel tank bracket

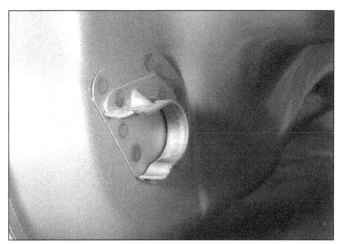

2.8 These cups on the tank fit over rubber dampers on the frame

Installation

Refer to illustration 2.8

7   Before installing the tank, check the condition of the rubber mounting dampers - if they're hardened, cracked, or show any other signs of deterioration, replace them **(see illustration 2.6)**.
8   When installing the tank, reverse the removal procedure. Make sure the metal cups seat properly on the front mounting dampers and do not pinch any control cables or wires **(see illustration)**.

3   Fuel tank - cleaning and repair

1   All repairs to the fuel tank should be carried out by a professional who has experience in this critical and potentially dangerous work. Even after cleaning and flushing of the fuel system, explosive fumes can remain and ignite during repair of the tank.
2   If the fuel tank is removed from the vehicle, it should not be placed in an area where sparks or open flames could ignite the fumes coming out of the tank. Be especially careful inside garages where a natural gas-type appliance is located, because the pilot light could cause an explosion.

4   Idle fuel/air mixture adjustment

Normal adjustment

Refer to illustration 4.3

1   Idle fuel/air mixture on these vehicles is preset at the factory and should not need adjustment unless the carburetor is overhauled or the pilot screw, which controls the mixture adjustment, is replaced.
2   The engine must be properly tuned up before making the adjustment (valve clearances set to specifications, spark plug in good condition and properly gapped).
3   To make an initial adjustment, turn the pilot screw clockwise until it seats lightly, then back it out the number of turns listed in this Chapter's Specifications **(see illustration)**. **Caution:** *Turn the screw just far enough to seat it lightly. If it's bottomed hard, the screw or its seat may be damaged, which will make accurate mixture adjustments impossible.*
4   Warm up the engine to normal operating temperature. Shut it off and connect a tachometer, following the tachometer manufacturer's instructions.
5   Restart the engine and compare idle speed to the value listed in the Chapter 1 Specifications. Adjust it if necessary.

4.3 The pilot screw (arrow) adjusts idle fuel/air mixture

6   If you're working on a 1997 or earlier model, turn the pilot screw clockwise just until engine speed drops or the engine starts to run roughly. Note how many turns were required. Turn the pilot screw counterclockwise, again until the engine starts to slow or run roughly, and note how many turns were required. Finally, turn the pilot screw clockwise to a point halfway between the first two points.
7   If you're working on a 1998 or later model, turn the pilot screw in or out from the initial setting to obtain the maximum idle speed. Reset the idle speed to Specifications with the throttle stop screw. Now, slowly turn the pilot screw in until engine speed drops 100 rpm, then turn it back out 3/4-turn.

High altitude adjustment

8   If the vehicle is normally used at altitudes from sea level to 5000 feet (1500 meters), use the normal main jet and pilot screw setting. If it's used at altitudes between 3000 and 8000 feet (1000 and 2500 meters), the main jet and pilot screw setting must be changed to compensate for the thinner air. **Caution:** *Don't use the vehicle for sustained operation below 5000 feet (15 meters) with the main jet and pilot screw at the high altitude settings or the engine may overheat and be damaged.*
9   Refer to Section 7 and change the main jet to the high altitude jet listed in this Chapter's Specifications.
10   Set the pilot screw to the standard setting, then turn it in the additional amount listed in this Chapter's Specifications.
11   With the vehicle at high altitude, refer to Chapter 1 and adjust the idle speed.

6.3a Loosen the clamping band on the connecting tube (arrow) . . .

6.3b . . . and on the intake manifold tube (arrow)

5   Carburetor overhaul - general information

1   Poor engine performance, hesitation, hard starting, stalling, flooding and backfiring are all signs that major carburetor maintenance may be required.
2   Keep in mind that many so-called carburetor problems are really not carburetor problems at all, but are ignition system malfunctions or mechanical problems within the engine. Try to establish for certain that the carburetor is in need of maintenance before beginning a major overhaul.
3   Check the fuel tap and its strainer screen, the in-tank fuel strainer, the fuel lines, the intake manifold clamps, the O-ring between the intake manifold and cylinder head, the vacuum hoses, the air filter element, the cylinder compression, the spark plug and the ignition timing before assuming that a carburetor overhaul is required. If the vehicle has been unused for more than a month, refer to Chapter 1, drain the float chamber and refill the tank with fresh fuel.
4   Most carburetor problems are caused by dirt particles, varnish and other deposits which build up in and block the fuel and air passages. Also, in time, gaskets and O-rings shrink or deteriorate and cause fuel and air leaks which lead to poor performance.
5   When the carburetor is overhauled, it is generally disassembled completely and the parts are cleaned thoroughly with a carburetor cleaning solvent and dried with filtered, unlubricated compressed air. The fuel and air passages are also blown through with compressed air to force out any dirt that may have been loosened but not removed by the solvent. Once the cleaning process is complete, the carburetor is reassembled using new gaskets, O-rings and, generally, a new inlet needle valve and seat.
6   Before disassembling the carburetors, make sure you have a carburetor rebuild kit (which will include all necessary O-rings and other parts), some carburetor cleaner, a supply of rags, some means of blowing out the carburetor passages and a clean place to work.

6   Carburetor - removal and installation

Warning: *Gasoline (petrol) is extremely flammable, so take extra precautions when you work on any part of the fuel system. Don't smoke or allow open flames or bare light bulbs near the work area, and don't work in a garage where a natural gas-type appliance (such as a water heater or clothes dryer) with a pilot light is present. Since gasoline is carcinogenic, wear latex gloves when there's a possibility of being exposed to fuel, and, if you spill any fuel on your skin, rinse it off immediately with soap and water. Mop up any spills immediately and do not store fuel-soaked rags where they could ignite. When you perform any kind of work on the fuel system, wear safety glasses and have a Class B type fire extinguisher on hand.*

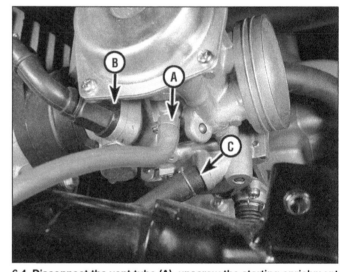

6.4 Disconnect the vent tube (A), unscrew the starting enrichment valve (B) and disconnect the fuel line (C)

Removal

Refer to illustrations 6.3a, 6.3b, 6.4, 6.7a and 6.7b
1   Remove the fuel tank (see Section 2).
2   Remove the fuel tank bracket **(see illustration 2.6).**
3   Loosen the clamping bands on the air cleaner duct and the intake manifold tube **(see illustrations)**. Work the carburetor free of the duct and tube and lift it up.
4   Disconnect the vent hose from the carburetor **(see illustration)**. Unscrew the starting enrichment valve nut and pull the valve out of the carburetor body. The upper end of the fuel line was disconnected as part of fuel tank removal, but if it you plan to overhaul the carburetor, disconnect the fuel line from the carburetor as well.
5   Loosen the throttle cables all the way and disconnect them from the throttle grip (see Section 10).
6   Refer to Section 10 and disconnect the throttle cable from the carburetor body. The carburetor can now be removed.
7   Check the intake manifold tube for cracks, deterioration or other damage **(see illustration)**. If it has visible defects, or if there's reason to suspect its O-ring is leaking, remove it and inspect the O-ring **(see illustration)**.
8   After the carburetor has been removed, stuff clean rags into the intake tube (or the intake port in the cylinder head, if the tube has been removed) to prevent the entry of dirt or other objects.

6.7a Remove the bolts (arrows) . . .

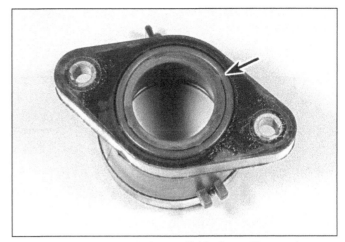

6.7b . . . then detach the manifold tube and inspect its O-ring (arrow)

7.2a Remove the four screws securing the vacuum chamber cover to the carburetor body (arrows)

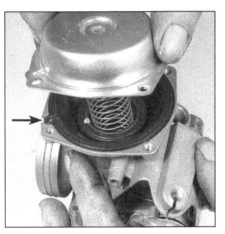

7.2b Lift the cover off and remove the piston spring; note the location of the tab (arrow) which fits into a notch

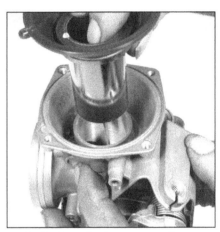

7.2c Peel the diaphragm away from its groove in the carburetor body, being careful not to tear it, and lift out the diaphragm/piston assembly

Installation

9   Refer to Sections 10 and 11 and connect the throttle and choke cables to the carburetor.
10   Connect the fuel hose to the carburetor (if it was disconnected) and connect the air vent tube **(see illustration 6.4)**.
11   Slip the clamping band onto the intake tube. Position the pin on the band in the intake tube groove. Position the carburetor in the intake tube and tighten the clamping band screw securely.
12   Adjust the throttle freeplay (see Chapter 1).
13   The remainder of installation is the reverse of the removal steps.

---

7   Carburetors - disassembly, cleaning and inspection

Warning: *Gasoline (petrol) is extremely flammable, so take extra precautions when you work on any part of the fuel system. Don't smoke or allow open flames or bare light bulbs near the work area, and don't work in a garage where a natural gas-type appliance (such as a water heater or clothes dryer) with a pilot light is present. Since gasoline is carcinogenic, wear latex gloves when there's a possibility of being exposed to fuel, and, if you spill any fuel on your skin, rinse it off immediately with soap and water. Mop up any spills immediately and do not store fuel-soaked rags where they could ignite. When you perform any kind of work on the fuel system, wear safety glasses and have a Class B type fire extinguisher on hand.*

7.2d Push in on the needle jet holder and turn it 90-degrees counterclockwise with an 8 mm socket . . .

Disassembly

Refer to illustrations 7.2a through 7.2s

1   Remove the carburetor as described in Section 6. Set it on a clean working surface.
2   Refer to the accompanying illustrations to disassemble the carburetor **(see illustrations)**.

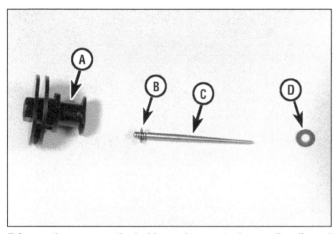

7.2e ... then remove the holder and separate the needle, clip and washer from the piston; standard position for the clip is in the third groove

a)   Holder   c)   Jet needle
b)   Clip   d)   Washer

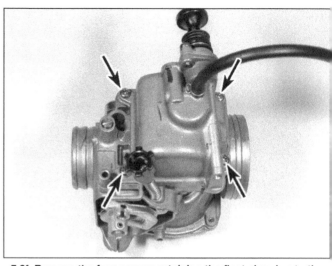

7.2f   Remove the four screws retaining the float chamber to the carburetor body (arrows) . . .

7.2g ... then detach the float chamber and remove its O-ring

7.2h   Remove the float chamber baffle, push the float pivot pin out and detach the float (and fuel inlet valve needle) from the carburetor body

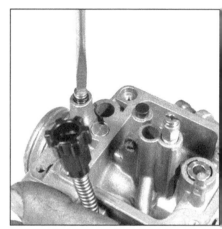

7.2i   Turn the pilot screw in, counting the number of turns until it bottoms lightly, and record the number for use when installing the screw . . .

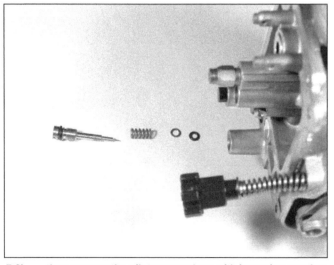

7.2j ... then remove the pilot screw along with its spring, washer and O-ring

7.2k   Unscrew the starter jet

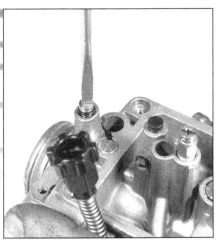

7.2l Unscrew the slow jet and pull it out . . .

7.2m . . . noting that the narrow end goes in first

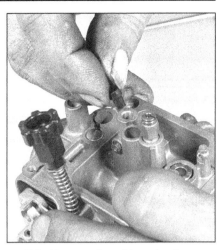

7.2n Remove the rubber plug from its passage

7.2o Unscrew the main jet from the needle jet holder . . .

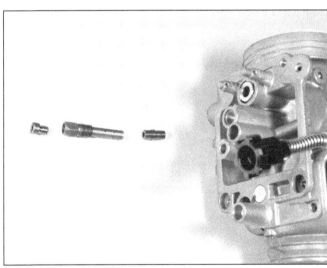

7.2p . . . then unscrew the needle jet holder and remove the needle jet

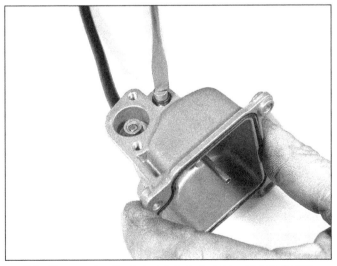

7.2q Remove the float chamber drain screw and its O-ring

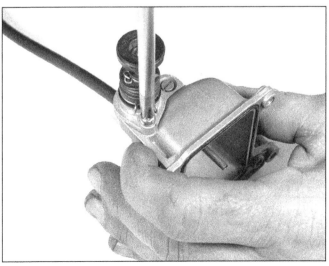

7.2r Remove the primer knob mounting screws . . .

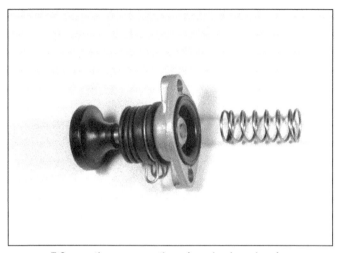

7.2s . . . then remove the primer knob and spring

7.15 Check the starting enrichment valve for wear and damage and replace it if necessary; on 1994 and later models apply a thin coat of multi-purpose grease to the end of the valve threads (arrow) on installation

Cleaning

Caution: *Use only a carburetor cleaning solution that is safe for use with plastic parts (be sure to read the label on the container).*

3   Submerge the metal components in the carburetor cleaner for approximately thirty minutes (or longer, if the directions recommend it).

4   After the carburetor has soaked long enough for the cleaner to loosen and dissolve most of the varnish and other deposits, use a brush to remove the stubborn deposits. Rinse it again, then dry it with compressed air. Blow out all of the fuel and air passages in the main and upper body. **Caution:** *Never clean the jets or passages with a piece of wire or a drill bit, as they will be enlarged, causing the fuel and air metering rates to be upset.*

Inspection

Refer to illustration 7.15

5   Check the operation of the primer knob. If it doesn't move smoothly, replace it, along with the return spring. If the rubber components of the knob are deteriorated or damaged, remove the clips, disassemble the knob and replace them.

6   Check the tapered portion of the pilot screw for wear or damage. Replace the pilot screw if necessary.

7   Check the carburetor body, float chamber and vacuum chamber cover for cracks, distorted sealing surfaces and other damage. If any defects are found, replace the faulty component, although replacement of the entire carburetor will probably be necessary (check with your parts supplier for the availability of separate components).

8   Check the jet needle for straightness by rolling it on a flat surface (such as a piece of glass). Replace it if it's bent or if the tip is worn.

9   Check the tip of the fuel inlet valve needle. If it has grooves or scratches in it, it must be replaced. Push in on the rod in the other end of the needle, then release it - if it doesn't spring back, replace the valve needle.

10   Check the O-rings on the float chamber and the drain plug (in the float chamber). Replace them if they're damaged.

11   Operate the throttle shaft to make sure the throttle butterfly valve opens and closes smoothly. If it doesn't, replace the carburetor.

12   Check the floats for damage. This will usually be apparent by the presence of fuel inside one of the floats. If the floats are damaged, they must be replaced.

13   Check the diaphragm for splits, holes and general deterioration. Holding it up to a light will help to reveal problems of this nature.

14   Insert the vacuum piston in the carburetor body and see that it moves up-and-down smoothly. Check the surface of the piston for wear. If it's worn excessively or doesn't move smoothly in the bore, replace the carburetor.

15   Check the starting enrichment valve for wear or damage and replace it if any defects are found **(see illustration)**.

8   Carburetors - reassembly and float height check

Caution: *When installing the jets, be careful not to over-tighten them - they're made of soft material and can strip or shear easily.*

Note: *When reassembling the carburetor, be sure to use the new O-rings, gaskets and other parts supplied in the rebuild kit.*

1   Install the clip on the jet needle if it was removed. Place it in the needle groove listed in this Chapter's Specifications. Install the washer, needle and clip in the piston, then install the holder, press it down and turn it 90-degrees clockwise with an 8 mm socket.

2   Install the pilot screw (if removed) along with its spring, washer and O-ring, turning it in until it seats lightly. Now, turn the screw out the number of turns that was previously recorded.

3   Install the diaphragm/vacuum piston assembly into the carburetor body. Lower the spring into the piston. Seat the bead of the diaphragm into the groove in the top of the carburetor body, making sure the diaphragm isn't distorted or kinked **(see illustration 7.2b)**. This is not always an easy task. If the diaphragm seems too large in diameter and doesn't want to seat in the groove, place the vacuum chamber cover over the carburetor diaphragm, insert your finger into the throat of the carburetor and push up on the vacuum piston, holding it almost all the way up. Push down gently on the vacuum chamber cover - it should drop into place, indicating the diaphragm has seated in its groove. Once this occurs, install at least two of the vacuum chamber cover screws before you let go of the piston.

4   Install the remaining vacuum chamber cover screws and tighten all of them securely.

5   Reverse the disassembly steps to install the jets and rubber plug.

6   Invert the carburetor. Attach the fuel inlet valve needle to the float. Set the float into position in the carburetor, making sure the valve needle seats correctly. Install the float pivot pin. To check the float height, hold the carburetor so the float hangs down, then tilt it back until the valve needle is just seated. Measure the distance from the float chamber gasket surface to the top of the float and compare your measurement to the float height listed in this Chapter's Specifications. There's no means of adjustment; if it isn't as specified, replace the float and needle valve.

7   Install the O-ring into the groove in the float chamber. Place the float chamber on the carburetor and install the screws, tightening them securely.

8   Install the spring and primer knob, taking care to locate the diaphragm correctly.

9.2 Loosen the clamping band (arrow) and detach the connecting tube from the carburetor

9.3 On later models, unclip the breather hose from the top of the intake duct (arrow) . . .

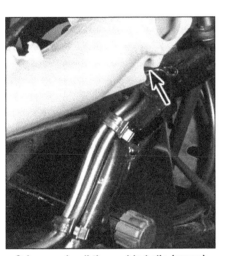

9.4 . . . and pull the molded clip (arrow) from the frame

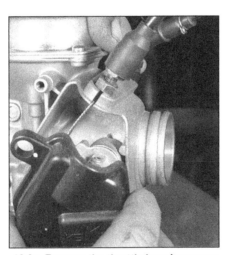

10.3a Remove the throttle housing cover screw and lift off the cover

10.3b Loosen the locknut and adjusting nut (arrows) and slip the cable out of the slot, then align the cable with the slot in the throttle pulley and slip it out

9   Air cleaner housing - removal and installation

Removal

Refer to illustrations 9.2, 9.3 and 9.4

1   Refer to Section 2 and remove the fuel tank and its bracket.

2   Loosen the clamp and detach the connecting tube from the carburetor **(see illustration)**.

3   Detach the crankcase breather tube from its clamp on the frame rail. On later models, unclip the tube from the top of the intake duct **(see illustration)**.

4   On 1988 through 1992 models, loosen the clamp and detach the front end of the intake duct from the frame bracket. On 1993 and later models, pull the molded clip portion of the intake duct off the frame bar **(see illustration)**.

5   Refer to *Air cleaner - filter element and drain tube cleaning* in Chapter 1 and remove the two air cleaner housing bolts. Lift the air cleaner housing out of the frame, together with the intake duct and carburetor connecting tube.

6   Installation is the reverse of the removal steps. On 1988 through 1992 models, position the front end of the intake duct so there is a gap

of 1 to 3 mm (3/64 to 1/8-inch) between the end of the intake duct and the welded bracket on the frame.

10   Throttle cable and housing - removal, installation and adjustment

1   The throttle on these vehicles is operated by a thumb lever on the right handlebar.

Removal

Refer to illustrations 10.3a, 10.3b, 10.4a, 10.4b and 10.6

2   Remove the fuel tank (see Section 2) and the carburetor (see Section 6).

3   Remove the throttle housing cover from the carburetor **(see illustration)**. Loosen the locknut and adjusting nut to create slack in the cable **(see illustration)**, then rotate the throttle pulley, align the cable with the pulley slot and slip the cable out of the pulley. Slip the cable out of the bracket slot to complete disconnection at the carburetor end.

4   Remove the cover from the throttle housing on the handlebar,

10.4a Bend back the lockwasher tab (arrow) and remove the nut, lockwasher and lever . . .

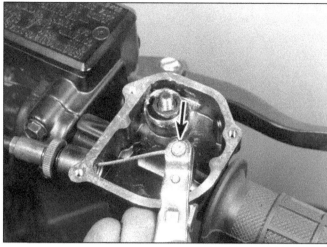

10.4b . . . then align the lever slot (arrow) with the cable and slip the cable out (be careful not to lose the spring)

10.6 Remove the clamp screws (arrows) to detach the throttle housing

10.7 Align the punch mark on the handlebar (under the throttle housing, right arrow) with the outer end of the throttle housing; align the throttle housing line with the brake master cylinder seam (left arrows)

remove the lever components and disconnect the cable **(see illustrations)**.

5   Remove the cable, noting how it's routed.

6   If necessary, remove the throttle housing clamp screws and detach the throttle housing from the handlebar **(see illustration)**.

Installation

Refer to illustration 10.7

7   If the throttle housing was removed, install it on the handlebar and tighten its clamp screws loosely. Position the housing so its outer end is aligned with the handlebar punch mark. Align the cast line on the housing with the seam of the brake master cylinder and clamp **(see illustration)**. Install the clamp and screws and tighten them securely.

8   Route the cable into place. Make sure it doesn't interfere with any other components and isn't kinked or bent sharply.

9   Lubricate the end of the cable with multi-purpose grease and connect it to the throttle pulley at the carburetor. Pass the inner cable through the slot in the bracket, then seat the cable housing in the bracket. Make sure all cable retainers are securely installed.

10   Reverse the disconnection steps to connect the throttle cable to the handlebar lever. Operate the lever and make sure it returns to the idle position by itself under spring pressure. **Warning:** *If the lever doesn't return by itself, find and solve the problem before continuing with installation. A stuck lever can lead to loss of control of the vehicle.*

Adjustment

11   Follow the procedure outlined in Chapter 1, *Throttle operation/lever freeplay - check and adjustment*, to adjust the cable.

12   Turn the handlebars back and forth to make sure the cables don't cause the steering to bind.

13   Once you're sure the cable operates properly, install the covers on the throttle housing and carburetor.

14   Install the fuel tank bracket and the tank.

15   With the engine idling, turn the handlebars through their full travel (full left lock to full right lock) and note whether idle speed increases. If it does, the cable is routed incorrectly. Correct this dangerous condition before riding the vehicle.

11 Choke cable - removal and installation

Removal

Refer to illustrations 11.3a, 11.3b and 11.3c

1   Remove the seat and fuel tank (see Chapter 7 and Section 2).

2   Unscrew the cable, together with the starting enrichment valve,

11.3a Remove the cable clip (arrow) . . .

11.3b . . . and slip the cable end out of the housing slot (arrow) . .
.

11.3c . . . and the lever slot (arrow)

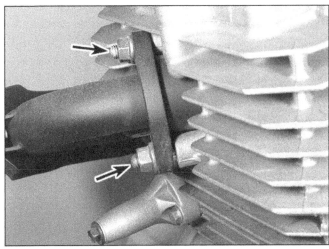

12.2 Remove the holder nuts (arrows) and slip the holder
off the studs

from the carburetor **(see illustration 6.4)**.

3   At the left handlebar, remove the clip that secures the choke
cable to the switch housing **(see illustration)**. Slip the cable out of the
slots in the housing and lever **(see illustrations)**.

4   Detach the cable from any clips and remove it, noting how it's
routed.

Installation

5   Installation is the reverse of the removal steps. Make sure the
cable is securely fastened in its clips.

6   Install the fuel tank and all of the other components that were
previously removed.

12 Exhaust system - removal and installation

Refer to illustrations 12.2, 12.3, 12.4a and 12.4b

Warning: *Wait until the exhaust system has cooled completely before
beginning this procedure.*

1   Refer to Chapter 7 and remove the right side cover.

2   Remove the exhaust pipe holder nuts and slide the holder off the
mounting studs **(see illustration)**.

3   If necessary, unbolt the heat shield and remove it from the
exhaust pipe **(see illustration)**.

12.3 Remove the heat shield bolts if necessary (two forward
bolts shown)

12.4a Remove the muffler bolt near the kickstarter . . .

12.4b . . . and the bolts above and below the muffler (arrows)

4   Remove the muffler mounting bolts **(see illustrations)**.

5   Pull the exhaust system forward, separate the pipe from the cylinder head and remove the system from the machine. Remove the gasket from the exhaust port.

6   Installation is the reverse of removal, with the following additions:

a)   *Be sure to install a new gasket in the cylinder head exhaust port.*

b)   *Tighten the muffler mounting bolts and heat shield bolts to the torque listed in this Chapter's Specifications. Tighten the holder nuts securely.*

Chapter 4
Ignition system

Contents

	Section			Section
Engine kill switch - check, removal and installation	See Chapter 8		Ignition (main) switch and key lock cylinder - check, removal and installation	See Chapter 8
General information	1		Ignition system - check	2
Ignition coil - check, removal and installation	3		Ignition timing - general information and check	6
Ignition control module (ICM) - harness check, removal and installation	5		Pulse generator - check, removal and installation	4
			Spark plug replacement	See Chapter 1

Specifications

Ignition coil resistance (at 20-degrees C/68-degrees F)

Primary resistance	0.1 to 0.2 ohms
Secondary resistance	
1988 through 1993 models	
With plug cap attached	8100 to 10,000 ohms
With plug cap detached	3600 to 4500 ohms
1994 and later models	
With plug cap attached	6500 to 9800 ohms
With plug cap detached	2700 to 3500 ohms

Pulse generator resistance

(at 20-degrees C/68-degrees F)	290 to 360 ohms

Ignition timing

At idle	13-degrees BTDC
At 4500 +/- 100 rpm	31-degrees BTDC

Torque specifications

Pulse generator screws or Allen bolts	6 Nm (48 inch-lbs)*

*Apply non-permanent thread locking agent to the bolt threads.

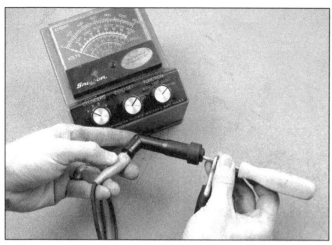

2.5 Unscrew the spark plug cap from the plug wire and measure its resistance with an ohmmeter

2.13 A simple spark gap testing fixture can be made from a block of wood, two nails, a large alligator clip, a screw and a piece of wire

1   General information

This vehicle is equipped with a battery operated, fully transistorized, breakerless ignition system. The system consists of the following components:

Pulse generator
Ignition control module (ICM)
Battery and fuse
Ignition coil
Spark plug
Engine kill (stop) and main (key) switches
Primary and secondary (HT) circuit wiring

The transistorized ignition system functions on the same principle as a DC ignition system with the pulse generator and ICM performing the tasks previously associated with the breaker points and mechanical advance system. As a result, adjustment and maintenance of ignition components is eliminated (with the exception of spark plug replacement).

Because of their nature, the individual ignition system components can be checked but not repaired. If ignition system troubles occur, and the faulty component can be isolated, the only cure for the problem is to replace the part with a new one. Keep in mind that most electrical parts, once purchased, can't be returned. To avoid unnecessary expense, make very sure the faulty component has been positively identified before buying a replacement part.

2   Ignition system - check

Refer to illustrations 2.5 and 2.13

Warning: *Because of the very high voltage generated by the ignition system, extreme care should be taken when these checks are performed.*

1   If the ignition system is the suspected cause of poor engine performance or failure to start, a number of checks can be made to isolate the problem.

2   Make sure the ignition kill (stop) switch is in the Run or On position.

Engine will not start

3   Refer to Chapter 1 and disconnect the spark plug wire. Connect the wire to a spare spark plug and lay the plug on the engine with the threads contacting the engine. If necessary, hold the spark plug with an insulated tool. Crank the engine over and make sure a well-defined, blue spark occurs between the spark plug electrodes. **Warning:** *Don't remove the spark plug from the engine to perform this check -*

atomized fuel being pumped out of the open spark plug hole could ignite, causing severe injury!

4   If no spark occurs, the following checks should be made:

5   Unscrew the spark plug cap from the plug wire and check the cap resistance with an ohmmeter **(see illustration)**. If the resistance is infinite, replace it with a new one.

6   Make sure all electrical connectors are clean and tight. Check all wires for shorts, opens and correct installation.

7   Check the battery voltage with a voltmeter. If the voltage is less than 12-volts, recharge the battery.

8   Check the ignition fuse and the fuse connections (see Chapter 8). If the fuse is blown, replace it with a new one; if the connections are loose or corroded, clean or repair them.

9   Refer to Section 3 and check the ignition coil primary and secondary resistance.

10   Refer to Section 4 and check the pulse generator resistance.

11   If the preceding checks produce positive results but there is still no spark at the plug, refer to Section 5 and check the ICM.

Engine starts but misfires

12   If the engine starts but misfires, make the following checks before deciding that the ignition system is at fault.

13   The ignition system must be able to produce a spark across a seven millimeter (1/4-inch) gap (minimum). A simple test fixture **(see illustration)** can be constructed to make sure the minimum spark gap can be jumped. Make sure the fixture electrodes are positioned seven millimeters apart.

14   Connect one of the spark plug wires to the protruding test fixture electrode, then attach the fixture's alligator clip to a good engine ground (earth).

15   Crank the engine over with the key in the On position and see if well-defined, blue sparks occur between the test fixture electrodes. If the minimum spark gap test is positive, the ignition coil is functioning properly. If the spark will not jump the gap, or if it is weak (orange colored), refer to Steps 5 through 11 of this Section and perform the component checks described.

3   Ignition coil - check, removal and installation

Check

Refer to illustration 3.4

1   In order to determine conclusively that the ignition coils are defective, they should be tested by an authorized Honda dealer service department which is equipped with the special electrical tester required for this check.

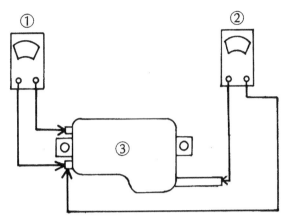

3.4 Ignition coil test

1   *Measure primary winding resistance*
2   *Measure secondary winding resistance*
3   *Ignition coil*

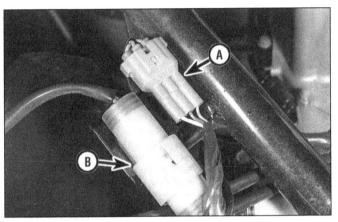

4.1 The pulse generator connector (A) and alternator connector (B) are on the left side of the frame

2   However, the coil can be checked visually (for cracks and other damage) and the primary and secondary coil resistances can be measured with an ohmmeter. If the coil is undamaged, and if the resistances are as specified, it is probably capable of proper operation.
3   To check the coil for physical damage, they must be removed (see Step 9). To check the resistance, remove the fuel tank (see Chapter 3), unplug the primary circuit electrical connectors from the coil and remove the spark plug wire from the spark plug. Mark the locations of all wires before disconnecting them **(see illustration)**.
4   To check the coil primary resistance, attach one ohmmeter lead to one of the primary terminals and the other ohmmeter lead to the other primary terminal **(see illustration)**.
5   Place the ohmmeter selector switch in the Rx1 position and compare the measured resistance to the value listed in this Chapter's Specifications.
6   If the coil primary resistance is as specified, check the coil secondary resistance by disconnecting the meter leads from the primary terminals and attaching them between the spark plug wire terminal and the green primary terminal **(see illustration 3.4)**.
7   Place the ohmmeter selector switch in the Rx100 position and compare the measured resistance to the values listed in this Chapter's Specifications.
8   If the resistances are not as specified, unscrew the spark plug cap from the plug wire and check the resistance between the green primary terminal and the end of the spark plug wire. If it is now within specifications, the spark plug cap is bad. If it's still not as specified, the coil is probably defective and should be replaced with a new one.

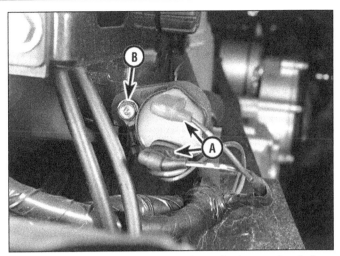

3.9 It's a good idea to mark the ignition coil primary wires before disconnecting them

A   *Primary wire terminals*
B   *Mounting bolt*

Removal and installation

Refer to illustration 3.9

9   To remove the coil, refer to Chapter 3 and remove the fuel tank, then disconnect the spark plug wire from the plug. After labeling them with tape to aid in reinstallation, unplug the coil primary circuit electrical connectors **(see illustration)**.
10   Support the coil with one hand and remove the coil mounting bolt, then lift the coil out.
11   Installation is the reverse of removal. Make sure the primary circuit electrical connectors are attached to the proper terminals. Just in case you forgot to mark the wires, the black/yellow wire connects to the black primary terminal and the green wire attaches to the green terminal.

---

4   Pulse generator - check, removal and installation

Check

Refer to illustrations 4.1 and 4.3

1   Locate the pulse generator connector on the left side of the vehicle, on the frame rail inside the rear fender **(see illustration)**. Disconnect the pulse generator connector.
2   Set the ohmmeter to R x 100. Measure the resistance between a good ground/earth and the blue/yellow wire terminal in the electrical connector. Compare the reading with the value listed in this Chapter's Specifications.
3   If the reading is not within the specified range, refer to Chapter 8 and remove the alternator cover from the left side of the crankcase. Disconnect the wire from the pulse generator, which is mounted inside the cover **(see illustration)**. Measure the resistance between a good ground/earth and the pulse generator terminal. If it's now within specifications, there's a break or bad connection in the wire. If it's still incorrect, replace the pulse generator.

Removal

4   Remove the alternator cover and disconnect the pulse generator connector, if you haven't already done so (see Chapter 8).
5   Check to see if the pulse generator wire is clipped into the plastic retainer cast in the side of the pulse generator **(see illustration 4.3)**. If it is, unclip it.
6   Unscrew the pulse generator screws or Allen bolts and remove the pulse generator.

4.3 Check to see if the wire is clipped into the retainer (A) on the side of the pulse generator, then remove the screws or Allen bolts (B)

5.2 The ignition control module (A) is under the front fender; the oil temperature alarm unit (B) is next to the ICM (1988 through 1997) or built into it (1998 and later)

Installation

7   Installation is the reverse of the removal procedure, with the following additions:

a) *If the wire was clipped into the holder on the side of the pulse generator, make sure there's at least 40 mm (1.6 inch) of wire protruding from the harness sleeve, then clip the wire into the new pulse generator. On later models, the wire is short and need not be clipped into the retainer.*

b) *Apply non-permanent thread locking agent to the pulse generator screws or Allen bolts and tighten them to the torque listed in this Chapter's Specifications.*

8   Refer to Chapter 8 and reinstall the alternator cover.

5   Ignition control module (ICM) - harness check, removal and installation

Harness check

Refer to illustration 5.2

1   Refer to Chapter 7 and remove the front fender.

2   Disconnect the electrical connector from the ICM **(see illustration)**.

3   Set an ohmmeter at Rx1 and connect it between the black/yellow and green wire terminals in the harness. It should give the same reading as for ignition coil primary resistance, listed in this Chapter's Specifications.

4   Set the ohmmeter at Rx100 and connect it between the blue/yellow and green wire terminals in the harness. It should give the same reading as for the pulse generator resistance, listed in this Chapter's Specifications.

5   Place the shift pedal in the Neutral position. Connect the ohmmeter and note the reading:

a) *1988 models: between the light green/red and green terminals; no continuity.*

b) *1989 through 1993 models: between the gray and green terminals; continuity (little or no resistance).*

c) *1994 and later models: between the light green and green terminals; no continuity.*

6   Disconnect the ohmmeter. Connect a voltmeter between the black/white and green terminals. With the ignition switch in the On position and the engine kill switch in the Run position, the voltmeter should indicate battery voltage (approximately 12-volts).

7   If the harness failed any of the preceding tests, check the affected wires for breaks or poor connections.

8   If the harness and all other system components tested good, the ICM may be defective. Before buying a new one, it's a good idea to substitute a known good ICM or have it tested by a Honda dealer.

6   Ignition timing - general information and check

General information

1   Ignition timing need be checked only if you're troubleshooting a problem such as loss of power. Since the ignition timing can't be adjusted and since none of the ignition system parts are subject to mechanical wear, there's no need for regular checks.

2   The ignition timing is checked with the engine running, both at idle and at a higher speed listed in this Chapter's Specifications. Inexpensive neon timing lights should be adequate in theory, but in practice may produce such dim pulses that the timing marks are hard to see. If possible, one of the more precise xenon timing lights should be used, powered by an external source of the appropriate voltage. **Note:** *Don't use the vehicle's own battery as an incorrect reading may result from stray impulses within the electrical system.*

Check

3   Warm the engine to normal operating temperature, make sure the transmission is in Neutral, then shut the engine off.

4   Refer to *Valve clearance - check and adjustment* in Chapter 1 and remove the timing window cap.

5   Connect the timing light and a tune-up tachometer to the engine, following manufacturer's instructions.

6   Start the engine. Make sure it idles at the speed listed in the Chapter 1 Specifications. Adjust if necessary.

7   Point the timing light into the timing window. At idle, the line next to the F mark on the alternator rotor should align with the notch at the top of the timing window.

8   Raise engine speed to 4,500 rpm. The notch at the top of the timing window should now be between the two advance timing marks on the alternator rotor.

9   If the timing is incorrect and all other ignition components have tested as good, the ICM may be defective. Have it tested by a Honda dealer.

10   When the check is complete, grease the timing window O-ring, then install the O-ring and cap and disconnect the test equipment.

Chapter 5
Frame, suspension and final drive

Contents

	Section
Axle housing - removal and installation	17
Front differential and driveshaft (4WD models) - removal, inspection and installation	12
Front drive side shaft (4WD models) removal, inspection and installation	13
Front driveaxle (4WD models) - boot replacement and CV joint overhaul	11
Front driveaxles (4WD models) - removal and installation	10
General information	1
Handlebars - removal, inspection and installation	2
Rear axle - removal, inspection and installation	16
Rear driveshaft - removal, inspection and installation	22
Rear final drive unit - removal, inspection and installation	18
Shock absorbers - disassembly, inspection and reassembly	5
Shock absorbers - removal and installation	4

	Section
Steering knuckle bearing and lower balljoint replacement	8
Steering knuckles - removal, inspection, and installation	7
Steering shaft - removal, inspection, bearing replacement and installation	3
Suspension - check	See Chapter 1
Suspension arms - removal, inspection, balljoint replacement and installation	9
Swingarm - removal and installation	20
Swingarm bearings - check	19
Swingarm bearings - replacement	21
Tie-rods - removal, inspection and installation	6
Transfer case (4WD models) - disassembly, inspection and reassembly	15
Transfer case (4WD models) - removal and installation	14

Specifications

Shock absorbers

Front spring length
  1988 through 1992
    2WD models
      Standard .... 280.0 to 286.0 mm (11.02 to 11.26 inches)
      Minimum .... 277.2 mm (10.91 inches)
    4WD models
      Standard .... 244.4 to 250.4 mm (9.62 to 9.86 inches)
      Minimum .... 241.9 mm (9.52 inches)
  1993 on
    2WD models
      Standard .... 216.9 mm (8.54 inches)
      Minimum .... 212.5 mm (8.37 inches)
    4WD models
      Standard .... 223.8 mm (8.81 inches)
      Minimum .... 219.3 mm (8.63 inches)
Rear spring length
  1988 through 1992
    2WD models
      Standard .... 251.3 to 257.3 mm (9.89 to 10.13 inches)
      Minimum .... 248.8 mm (9.80 inches)
    4WD models
      Standard .... 253.0 to 259.0 mm (9.96 to 10.20 inches)
      Minimum .... 250.4 mm (9.86 inches)
  1993 on
    2WD models
      Standard .... 241.6 mm (9.51 inches)
      Minimum .... 236.7 mm (9.32 inches)
    4WD models
      Standard .... 243.3 mm (9.58 inches)
      Minimum .... 238.4 mm (9.39 inches)

Shock absorbers (continued)

Tie-rod balljoint spacing
    2WD models
        1988 through 1992.. 300 mm (11.8 inches)
        1993 on .... 345.5 mm (13.6 inches)
    4WD models... 343 mm (13.5 inches)
Rear axle runout (maximum)... 3.0 mm (0.12 inch)

Torque specifications

Handlebars
    Grip end bolts... 10 Nm (84 in-lbs)
    Handlebar bracket bolts.. 27 Nm (20 ft-lbs)
    Handlebar bracket nuts... 40 Nm (29 ft-lbs) (1)
Tie-rod nuts .... 55 Nm (40 ft-lbs)
Tie-rod locknuts... 55 Nm (40 ft-lbs)
Steering shaft nut
    1988 through 1992
        2WD models .... 70 Nm (51 ft-lbs)
        4WD models .... 70 Nm (51 ft-lbs) (2)
    1993 on .... 110 Nm 79 ft-lbs)
Steering shaft upper bracket bolts .... 33 Nm (24 ft-lbs)
Front shock absorbers
    1988 through 1992
        2WD models
            Balljoint... 38 Nm (27 ft-lbs)
            Upper nut .... 55 Nm (40 ft-lbs)
            Lower pinch bolt.. 55 Nm (40 ft-lbs)
        4WD models
            Upper joint... 38 Nm (27 ft-lbs)
            Mounting bolts and nuts... 25 Nm (18 ft-lbs) (1)
    1993 and later models
        Upper mounting bolt and nut....................................... 31 Nm (22 ft-lbs) (1)
        Lower mounting bolt and nut....................................... 45 Nm (33 ft-lbs) (1)
Rear shock absorbers
    1988 through 1992 models
        Upper bolt/nut.. 45 Nm (33 ft-lbs) (1)
        Lower bolt/nut.. 35 Nm (25 ft-lbs) (1)
    1993 and later models (upper and lower)........................... 45 Nm (33 ft-lbs) (1)
Front suspension arm pivot bolts
    1988 through 1992 models .... 45 Nm (33 ft-lbs) (1)
    1993 and later models... 31 Nm (22 ft-lbs) (1)
Balljoint nuts
    1988 through 1992 2WD models .... 50 to 60 Nm (36 to 43 ft-lbs)
    1993 and later 2WD and all 4WD models .... 30 to 36 Nm (22 to 26 ft-lbs)
Brake hose and breather tube retainer bolts
    2WD models... 22 Nm (16 ft-lbs)
    4WD models... 12 Nm (108 in-lbs)
Front differential
    Front bracket bolts.. 20 Nm (14 ft-lbs)
    Mounting bolts/nuts .... 45 Nm (33 ft-lbs)
Front driveshaft cover bolts.. 10 Nm (84 in-lbs)
Transfer case
    Left mounting bolts .... 25 Nm (18 ft-lbs)
    Right mounting bolts .... 12 Nm (108 in-lbs)
    Cover bolts... 12 Nm (108 in-lbs)
    Drain bolt.. see Chapter 1
Front drive side shaft cover bolts .... 10 Nm (84 in-lbs)
Axle inner locknuts .... 40 Nm (29 ft-lbs)
Axle outer locknuts .... 130 Nm (94 ft-lbs) (3)
Rear axle housing bolts .... 50 Nm (36 ft-lbs)
Rear final drive unit mounting nuts .... 45 Nm (33 ft-lbs) (1)
Swingarm
    Left pivot bolt .... 115 Nm (83 ft-lbs)
    Right pivot bolt .... 4 Nm (36 in-lbs)
    Right pivot bolt locknut .... 115 Nm (83 ft-lbs)

1) Use new nuts; don't reuse the old ones.
2) Lubricate the threads and the flange of the nut with grease.
3) Apply non-permanent thread locking agent to the threads.

2.3 The grips are secured to the handlebars by a bolt and adhesive

2.4 Lift up the trim piece and remove the screws (arrows)

2.5 Unbolt the brackets to free the handlebar

1   General information

The front suspension on 1988 through 1992 2WD models consists of a single lower control arm and steering knuckle on each side of the vehicle, with the machine supported by "strut-type" shock absorbers with concentric coil springs. On all other models, there's an upper and lower control arm and steering knuckle on each side of the vehicle. A shock absorber with concentric coil spring is installed between each upper suspension arm and the frame.

The rear suspension on all models consists of a single shock absorber with concentric coil spring and a steel swingarm. Final drive is by a shaft, which passes through an integral tube on the swingarm.

The steering system consists of knuckles mounted at the outer ends of the front suspension and connected to a steering shaft by tie rods. The steering shaft is turned directly by handlebars.

2   Handlebars - removal, inspection and installation

1   The handlebars are a one-piece tube. The tube fits into a bracket, which is integral with the steering shaft. If the handlebars must be removed for access to other components, such as the steering shaft, simply remove the bolts and slip the handlebars off the bracket. It's not necessary to disconnect the cables, wires or brake hose, but it is a good idea to support the assembly with a piece of wire or rope, to avoid unnecessary strain on the cables, wires and the brake hose.

2   If the handlebars are to be removed completely, refer to Chapter 3 for the throttle housing removal procedure, Chapter 6 for the master cylinder removal procedure and Chapter 8 for the switch removal procedure.

Removal

Refer to illustrations 2.3, 2.4, 2.5 and 2.6

3   If you plan to remove one of the grips, remove its bolt while the handlebars are still bolted to the bracket **(see illustration)**.

4   Lift the trim cover and remove the handlebar cover screws **(see illustration)**.

5   Remove the handlebar bracket bolts and lift off the brackets **(see illustration)**. Lift the handlebar out of the bracket.

6   To remove the brackets, remove their mounting nuts and washers from below **(see illustration)**. Discard the nuts and use new ones on installation.

Inspection

7   Check the handlebar and brackets for cracks and distortion and replace them if any undesirable conditions are found.

Installation

Refer to illustration 2.8

8   Installation is the reverse of the removal steps, with the following additions:

   a) *If the bracket nuts were removed, install new nuts and tighten them to the torque listed in this Chapter's Specifications.*

   b) *When installing the handlebar to the brackets, line up the punch mark on the handlebar with the bracket seam* **(see illustration)**. *Tighten the bracket bolts to the torques listed in this Chapter's Specifications.*

2.6 Remove the bracket nuts and washers (arrows); discard the nuts and use new ones on installation

2.8 Line up the punch mark on the handlebar (arrow) with the bracket seam

3.3a Remove the bolts (arrows) and lift off the bracket . . .

3.3b . . . to expose the bushing; the Up mark and arrow must be upright when the bushing is installed

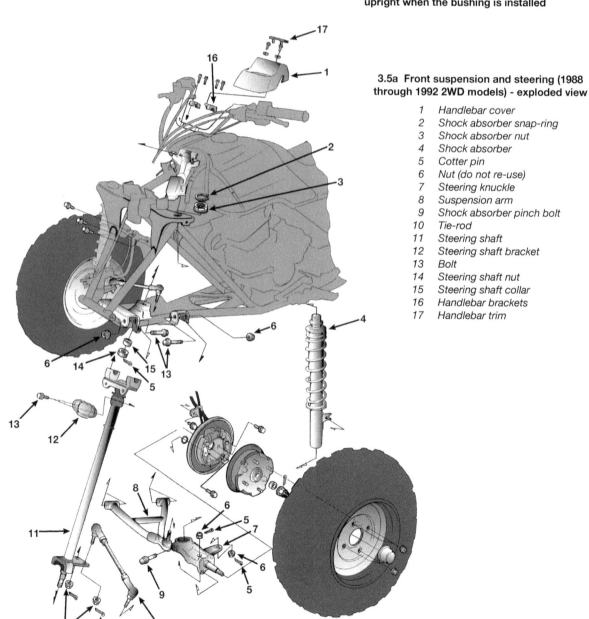

3.5a Front suspension and steering (1988 through 1992 2WD models) - exploded view

1   Handlebar cover
2   Shock absorber snap-ring
3   Shock absorber nut
4   Shock absorber
5   Cotter pin
6   Nut (do not re-use)
7   Steering knuckle
8   Suspension arm
9   Shock absorber pinch bolt
10   Tie-rod
11   Steering shaft
12   Steering shaft bracket
13   Bolt
14   Steering shaft nut
15   Steering shaft collar
16   Handlebar brackets
17   Handlebar trim

2125-5-3.5a HAYNES

3.5b Remove the cotter pin and nut, then detach the
steering arm from the shaft

3.12 Pry out the seals above and below the bearing
(upper seal shown) . . .

3.13 . . . and remove the snap-ring, then the bearing

3.15 Press the seals in; finger pressure should be enough, but if
not, tap the seals in with a socket the same diameter as the seal

3 Steering shaft - removal, inspection, bearing replacement and installation

Removal

Refer to illustrations 3.3a, 3.3b, 3.5a and 3.5b

1   Remove the handlebars and handlebar bracket (see Section 2).
2   Remove the front fender (see Chapter 7).
3   Unbolt the steering shaft bracket from the frame to expose the bushing **(see illustrations)**.
4   Refer to Section 6 and disconnect the inner ends of the tie-rods.
5   Remove the cotter pin and steering shaft nut **(see illustrations)**. If you're working on a 1993 and later 2WD model or any 4WD model, remove the steering arm from the shaft.
6   Remove the steering shaft from the vehicle.

Inspection

7   Clean all the parts with solvent and dry them thoroughly, using compressed air, if available.
8   Check the steering shaft bushing for wear, deterioration or damage **(see illustration 3.3b)**. Replace it if there's any doubt about its condition.
9   Check the steering shaft for bending or other signs of damage. On 1993 and later 2WD models and all 4WD models, inspect the

steering arm as well. Do not attempt to repair any steering components. Replace them with new parts if defects are found.
10   Insert a finger into the steering shaft bearing and turn the inner race. If it's rough, loose or noisy, replace it as described below.

Bearing replacement

Refer to illustrations 3.12, 3.13 and 3.15

11   If you're working on a 1988 through 1992 2WD model, remove the metal collar from the underside of the bearing **(see illustration 3.5a)**.
12   Pry the upper and lower grease seals out of the bearing housing in the frame **(see illustration)**.
13   Remove the snap-ring from on top of the bearing **(see illustration)**, then remove the bearing through the top of the housing.
14   Pack a new bearing with high-quality grease (preferably a moly-base grease). Position the bearing in the housing with its sealed side up and tap it into place with a bearing driver or socket that bears against the bearing outer race.
15   Install the snap-ring. Coat the lips of new upper and lower grease seals with grease, then install them in the housing **(see illustration)**.

Installation

Refer to illustration 3.16

16   Installation is the reverse of removal, with the following additions:

a) *Pack the steering shaft bushing's internal cavities with grease. Install it with its UP mark up* **(see illustration 3.3b)**.

b) *On all except 1989 through 1992 2WD models, align the wide splines on the steering arm and shaft, then install the steering arm* **(see illustration).**

c) *Lubricate the threads and flange of the steering shaft nut with grease.*

d) *Use new cotter pins and tighten all fasteners to the torque listed in this Chapter's Specifications.*

***

4   Shock absorbers - removal and installation

Front shock absorber

Removal

1988 through 1992 2WD models

1   Remove the front fender (see Chapter 7).

2   Securely block both rear wheels so the vehicle won't roll. Refer to Chapter 6 and remove the front wheels.

3   Loosen the pinch bolt at the lower end of the shock absorber (on the steering knuckle) and push the suspension arm down to disengage it from the shock **(see illustration 3.5a).**

4   At the upper end of the shock, remove the snap-ring and unscrew the nut. Lower the shock out of the frame hole and remove it from the vehicle.

1993 and later 2WD models; all 4WD models

Refer to illustration 4.6

5   Securely block both rear wheels so the vehicle won't roll. Refer to Chapter 6 and remove the front wheels.

6   Remove the mounting nut and bolt from the bottom of the shock, then from the top **(see illustration)**. Lift the shock out. Discard the mounting nuts and use new ones on installation.

Installation

1988 through 1992 2WD models

7   Slip the upper end of the shock into its socket on the frame. Install the nut and tighten it to the torque listed in this Chapter's Specifications. Position the snap-ring securely in its groove. **Caution:** *Use a new snap-ring if there's any doubt about its condition.*

8   Rotate the shock absorber so the projection on its lower end is aligned with the slit in the steering knuckle. Fit the lower end of the shock into the steering knuckle, install the pinch bolt and tighten it to the torque listed in this Chapter's Specifications.

1993 and later 2WD models; all 4WD models

9   Installation is the reverse of the removal steps. Use new nuts (don't re-use the old ones) and tighten the nuts and bolts to the torque listed in this Chapter's Specifications.

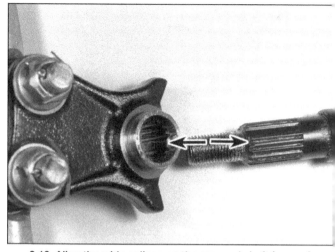

3.16 **Align the wide splines on the arm and shaft (arrows) when installing the arm**

Rear shock absorber

Removal

Refer to illustrations 4.11a and 4.11b

10   Securely block both front wheels so the vehicle can't toll. Jack up the rear end and support it securely with the rear wheels off the ground.

11   Remove the mounting bolts and nuts at the bottom of the shock, then at the top **(see illustrations)**. Lift the shock out of the vehicle. Discard the nuts and use new ones on installation.

Installation

12   Installation is the reverse of the removal steps. Use new nuts. Tighten the nuts and bolts to the torque listed in this Chapter's Specifications.

***

5   Shock absorbers - disassembly, inspection and reassembly

Warning: *Before overhauling the shocks, read through the procedure, paying special attention to the steps involved in compressing the spring. If you don't have access to the special tools needed on all except 1988 through 1992 2WD models, take the shock(s) to a Honda dealer or other shop equipped with the necessary special tools.*

4.6 **Remove the nut and bolt from the bottom of the shock absorber (shown), then from the top; use new nuts on installation and position the bolt heads facing forward**

4.11a **Remove the nut and bolt from the bottom of the shock absorber . . .**

4.11b **. . . then from the top**

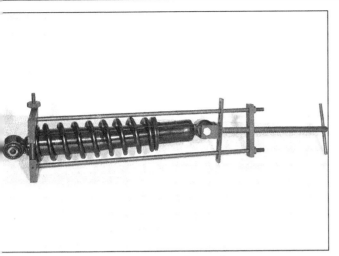

5.20 Place the shock absorber in a spring compressor

5.21 Compress the spring enough to remove the spring seat stopper

Front shock absorbers

1988 through 1992 2WD models

1   Compress the spring by hand, then remove the spring seat and take off the spring.

2   Pivot the balljoint at the upper end of the shock, Check it for looseness or rough movement and check the rubber boot for damage or deterioration. If any of these conditions are found, bend back the lockwasher, hold the shock absorber rod with a wrench on its flats and unscrew the balljoint from the rod.

3   Remove the lockwasher and seat guide from the rod, then slip the spring and spring guide off the shock absorber body.

4   Measure the free length of the spring. If its less than the value listed in this Chapter's Specifications, replace it.

5   Check the shock absorber for oil leaks where the rod meets the body. Check the rod for wear or damage and replace it if any problems are found.

6   Check the spring guide and seat guide for wear or damage and replace them as necessary.

7   Install the spring guide on the shock absorber body.

8   Install the seat guide, then install a new lockwasher with its tabs in the seat guide slots.

9   Thread the balljoint into the rod. Hold the rod with a wrench on its flats and tighten the balljoint to the torque listed in this Chapter's Specifications.

10   Bend the lockwasher up against the smaller hex on the balljoint with a hammer and punch.

11   Install the spring so its tightly wound coils will be up when the shock is installed.

12   Compress the spring by hand and install the spring seat.

1988 through 1992 4WD models

13   Place the shock absorber in a spring compressor (Honda tool 07967-KC10100 and 07GME-0010000 or 07959-3290001 with 07GME-0010100). The purpose of the compressor is to compress the spring safely, so it can't fly out and cause injury.

14   Compress the spring so the upper joint locknut is visible. Loosen the locknut and unscrew the upper joint from the shock absorber rod.

15   Release the compressor. Remove the spring guide and the spring.

16   Refer to Steps 4 through 6 above to inspect the shock. In addition, check the upper joint or wear or damage and replace it as necessary.

17   Install the spring and guide on the shock absorber. Position the spring so its tightly wound coils will be up when the shock is installed.

18   Compress the spring. Apply non-permanent thread locking agent to the damper rod threads and screw the upper joint locknut on.

19   With the locknut seated against the bottom thread of the shock rod, screw on the upper joint. Hold the locknut with a wrench and tighten the upper joint to the torque listed in this Chapter's Specifications.

Front shock absorbers (1993 and later models) and all rear shocks

Refer to illustrations 5.20, 5.21 and 5.22

20   Place the shock absorber in a spring compressor (Honda tool 07959-3290001) **(see illustration)**. The purpose of the compressor is to compress the spring safely, so it can't fly out and cause injury.

21   Compress the spring just enough so the spring stopper can be removed **(see illustration)**.

22   Release the spring compressor and take off the spring guide, spring and washer **(see illustration)**.

23   Refer to Steps 4 and 5 above and inspect the shock absorber. Check the spring seat stopper and washer for wear and damage and replace them as necessary.

24   Install the washer, spring and guide on the shock absorber. Position the spring so its tightly wound coils will be up when the shock is installed.

25   Compress the spring with the special tool and install the spring seat stopper. Release the spring and make sure the stopper holds it securely.

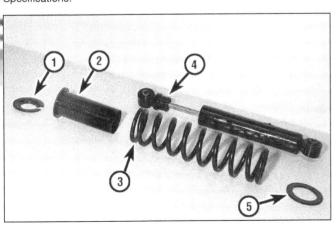

5.22 Front shock absorber details (1993 and later models)

1	Spring seat stopper	4	Shock absorber
2	Spring guide	5	Washer
3	Spring		

6.1 Straighten the cotter pin and pull it
out of the tie-rod nut . . .

6.2 . . . prevent the stud from turning with
an open-end wrench and unscrew the nut,
then detach the stud from
the steering knuckle

6.3 Remove the cotter pins and nuts
(arrows) and detach the inner ends of the
tie-rods from the steering arm

6   Tie-rods - removal, inspection and installation

Removal

Refer to illustrations 6.1, 6.2 and 6.3

1   Remove the cotter pin from the nut at the outer end of the tie-rod **(see illustration)**.
2   Hold the tie-rod flat with a wrench and undo the nut **(see illustration)**. Separate the tie-rod stud from the knuckle.
3   Repeat Steps 1 and 2 to disconnect the inner end of the tie rod **(see illustration)**. Discard the nuts and use new ones on installation.

Inspection

Refer to illustration 6.5

4   Check the tie-rod shaft for bending or other damage and replace it if any problems are found. Don't try to straighten the shaft.
5   Check the balljoint boot for cracks or deterioration **(see illustration)**. Twist and rotate the threaded stud. It should move easily, without roughness or looseness. If the boot or stud show any problems, unscrew the balljoint from the tie-rod and install a new one. **Note:** *The silver-colored locknut and balljoint labeled "L" go on the end of the tie-rod nearest the wrench flats. The gold-colored locknut and unlabeled balljoint go on the other end.*

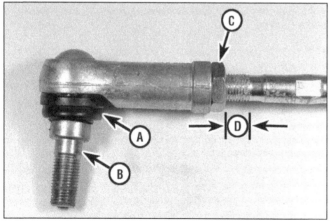

6.5 Check the boot for cracks or deterioration and make sure the
stud moves easily, without roughness or looseness

A)   Boot
B)   Stud
C)   Locknut

D)   Distance from locknut to
end of threads

Installation

6   Measure the distance from each locknut to the end of the threads and compare it with the value listed in this Chapter's Specifications **(see illustration 6.5)**. If it's incorrect, loosen the locknut and reposition the locknut and balljoint on the threads.
7   If you're working on a 1988 through 1992 2WD model, position the balljoint at each end of the tie-rod so the studs face in the same direction. On all other models, position the studs facing 180-degrees away from each other.
8   The remainder of installation is the reverse of the removal steps with the following additions:

a)   *Use new balljoint nuts (don't re-use the old ones). Tighten the new nuts to the torque listed in this Chapter's Specifications.*
b)   *Use new cotter pins and bend them to hold the nuts securely.*
c)   *Check front wheel toe-in and adjust as necessary (see Chapter 1).*

7   Steering knuckles - removal, inspection, and installation

Removal

Refer to illustration 7.1

1   Separating the balljoint(s) from the knuckle requires either a special Honda separator tool or its equivalent. Equivalent automotive tools can be rented, but they must be small enough for use on these vehicles. If the correct special tools aren't available, the knuckle can be removed as an assembly with the suspension arm(s) **(see illustration)**. This assembly can then be taken to a Honda dealer for balljoint removal (and knuckle bearing replacement on 4WD models).
2   Securely block both rear wheels so the vehicle won't roll. Loosen the front wheel nuts with the tires still on the ground, then jack up the front end, support it securely on jackstands and remove the front wheels.
3   Refer to Section 7 and disconnect the outer end of the tie-rod from the knuckle.
4   Remove the front brake panel (see Chapter 6). The front brake hose can be left connected, but be careful not to twist it and be sure to support the panel with wire or rope so it doesn't hang by the brake hose.

1988 through 1992 2WD models

Refer to illustration 7.6

5   Remove the cotter pin and nut from the balljoint and remove the pinch bolt that secures the lower end of the shock absorber to the knuckle **(see illustration 3.5a)**.
6   Detach the balljoint from the knuckle with a balljoint puller (Honda tool 07934-5510000 or equivalent) **(see illustration)**. A small two-jaw puller may also work.

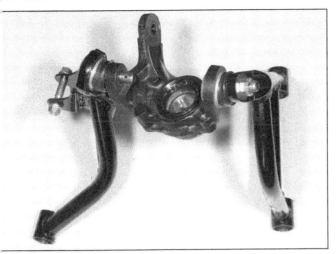

7.1 If you don't have the necessary balljoint separator tool, the suspension arms and knuckle can be removed as a unit and taken to a Honda dealer for balljoint separation (4WD model shown; others similar)

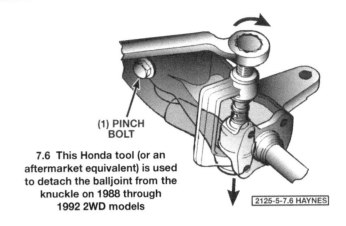

(1) PINCH BOLT

7.6 This Honda tool (or an aftermarket equivalent) is used to detach the balljoint from the knuckle on 1988 through 1992 2WD models

2125-5-7.6 HAYNES

7.7 Remove the cotter pins and nuts (arrows) . . .

1993 or later 2WD models and all 4WD models

Refer to illustrations 7.7 and 7.8

7   Remove the cotter pin and nut from the upper and lower balljoints **(see illustration)**.
8   Separate the balljoints from the knuckle with Honda tool 07MAC-SL00200 or 07941-6920003 or equivalent **(see illustration)**. Lubricate the puller jaw and the pressure bolt threads with multi-purpose grease. Slip the tool into position, taking care not to damage the rubber boot, and turn the adjusting bolt so the jaws are parallel. Tighten the pressure bolt by hand, make sure the jaws are still parallel (readjust the pressure bolt if necessary), then tighten the pressure bolt with a wrench until the balljoint stud pops out of the knuckle.

Inspection

Refer to illustration 7.11

9   Check the knuckle carefully for cracks, bending or other damage. Replace it if any problems are found. If the vehicle has been in a collision or has been bottomed hard, it's a good idea to have the knuckle magnafluxed by a machine shop to check for hidden cracks.
10   Check the balljoint boot for cracks or deterioration. Twist and rotate the threaded stud. It should move easily, without roughness or looseness. If the boot or stud show any problems, refer to Section 8 or Section 9 and replace the balljoint.

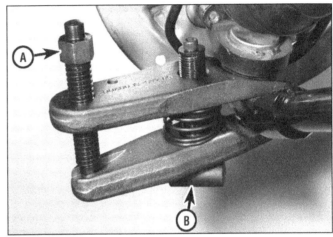

7.8 . . . and detach the balljoint from the knuckle with the special tool

A) Pressure bolt   B) Adjusting bolt

7.11 Turn the bearing inner race (A) and check for loose, rough or noisy movement; check the grease seals (B) for wear or damage

11   If you're working on a 4WD model, turn the knuckle bearing inner race with your finger **(see illustration)**. If it's rough, loose or noisy, refer to Section 8 and replace it. Replace the bearing grease seals if they show signs of leakage or wear.

Installation

12   Installation is the reverse of the removal steps, with the following additions: Use new cotter pins and tighten the balljoint nut(s) to the torque listed in this Chapter's Specifications.

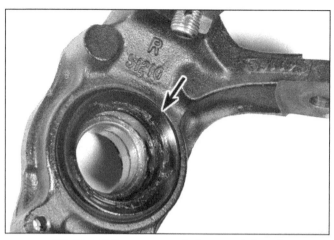

8.1 Pry out the grease seals (inner seal shown) before replacing the bearing

9.6 Remove the nuts and bolts (arrows); use new nuts on installation and install the bolts with their heads facing forward

8   Steering knuckle bearing and lower balljoint replacement

Bearing replacement

Refer to illustration 8.1

1   Pry out the bearing grease seals **(see illustration 7.11 and the accompanying illustration).**

2   Remove the bearing snap-ring from the outer side of the knuckle. Drive the bearing out of the knuckle with a bearing driver or socket that bears against the bearing outer race.

3   Pack a new bearing with grease, then drive it in with the same tool used for removal. Seat the bearing securely, then install the snap-ring and make sure it fits completely into its groove.

4   Tap in new grease seals with a bearing driver or socket the same diameter as the seals. Make sure the seals seat squarely in their bores, then lubricate the seal lips with grease.

Balljoint replacement

Note: *This procedure requires a press and special support tools (Honda 07946-187-0100 or equivalent). If you don't have them, take the knuckle to a Honda dealer for balljoint replacement.*

5   Remove the balljoint snap-ring.

6   Place the knuckle in a press with the balljoint stud up. Place the driver portion of the removal tool under the balljoint and the hollow portion over the stud and press the balljoint out.

7   Place the new balljoint on the press plate and place the knuckle over the stud.

8   Place the support tool over the stud so the stud will fit into the tool when the balljoint is pressed on. Press the balljoint into the knuckle; if it won't go easily, stop and make sure the support tool is aligned correctly.

9   Install the snap-ring and make sure it seats completely in its groove.

9   Suspension arms - removal, inspection, balljoint replacement and installation

Note: *This procedure describes removal and installation of the lower (and upper, on models so equipped) suspension arms. If you plan to remove only the upper or lower arm, ignore the steps which don't apply.*

Removal

1   Securely block both rear wheels so the vehicle won't roll. Loosen the front wheel nuts with the tires still on the ground, then jack up the front end, support it securely on jackstands and remove the front wheels.

2   Refer to Section 7 and remove the steering knuckle. If you're planning to remove the lower suspension arm on 1993 and later 2WD models or any 4WD model, remove the front bumper (see Chapter 7).

1988 through 1992 2WD models

3   Remove the pivot bolts and nuts and take the suspension arm out **(see illustration 3.5a).**

1993 and later 2WD models and all 4WD models

Refer to illustration 9.6

4   Unbolt the bottom end of the shock absorber from the upper suspension arm (see Section 5).

5   Detach the brake fluid hose and breather hose from the upper suspension arm.

6   Remove the nuts and bolts at the inner end of the suspension arm **(see illustration)** and pull the suspension arms out. Discard the nuts and use new ones on installation.

Inspection

Refer to illustration 9.8

7   Check the suspension arm(s) for bending, cracks or other damage. Replace damaged parts. Don't attempt to straighten them.

8   Check the rubber bushings at the inner end of the suspension arm for cracks, deterioration or wear of the metal insert **(see illustration).** Check the pivot bolts for wear as well. Replace the suspension arm if any problems are visible.

Balljoint replacement

9   Check the balljoint boot for cracks or deterioration. Twist and rotate the threaded stud. It should move easily, without roughness or looseness.

10   The balljoints on 1988 through 1992 2WD models can't be

9.8 Replace the suspension arms if the bushings are worn or damaged

10.3 The suspension arms and knuckle can be removed together if you don't have a balljoint separator tool

10.4 Pull on the joint (not on the shaft) to free the circlip (arrow) from the differential

replaced separately from the swingarm. If the boot or stud show any problems, replace the lower suspension arm.

11 The upper and lower balljoints on 1993 and later 2WD models, and the lower balljoint on all 4WD models, can be replaced with the methods and tools described in Section 8.

Installation

12 Installation is the reverse of the removal steps, with the following additions:

a) Use new nuts on the suspension arm pivot bolts.
b) Tighten the nuts and bolts slightly while the vehicle is jacked up, then tighten them to the torque listed in this Chapter's Specifications while the vehicle's weight is resting on the wheels.

10 Front driveaxles (4WD models) - removal and installation

Removal

Refer to illustrations 10.3 and 10.4

1 Securely block both rear wheels so the vehicle won't roll. Loosen the front wheel nuts with the tires still on the ground, then jack up the front end, support it securely on jackstands and remove the front wheels.

2 Remove the front panel (see Chapter 6).

3 Freeing the outer end of the driveaxle from the knuckle requires that the knuckle be pulled out off of the outer end of the driveaxle. If you have the necessary balljoint separator tool, you can use it to separate the balljoint studs from the knuckle, which will free the

knuckle to be pulled out **(see illustration 7.8)**. If you don't have the balljoint tool, the inner ends of the suspension arms can be unbolted (see Section 9) and the suspension arms and knuckle pulled out as a unit **(see illustration)**.

4 The inner end of the driveaxle is held in the final drive unit by a circlip **(see illustration)**. With the outer end of the driveaxle free of the knuckle, grasp the joint at the inner end firmly so it won't be pulled apart, then pull the driveaxle out of the final drive unit. **Caution:** *Pull the driveaxle straight out (don't let it tilt up, down or sideways) to prevent damage to the oil seal in the final drive unit.*

Installation

5 Installation is the reverse of the removal steps. After installing the driveaxle in the final drive unit, tug out on it to make sure the circlip is locked in place.

11 Front driveaxle (4WD models) - boot replacement and CV joint overhaul

Inner CV joint and boot

Disassembly

Refer to illustrations 11.3, 11.4, 11.6 and 11.7

1 Remove the driveaxle from the vehicle (see Section 10).

2 Mount the driveaxle in a vise. The jaws of the vise should be lined with wood or rags to prevent damage to the axleshaft.

3 Pry the boot clamp retaining tabs up with a small screwdriver and slide the clamps off the boot **(see illustration)**.

4 Slide the boot back on the axleshaft and pry the wire ring ball retainer from the outer race **(see illustration)**.

11.3 Pry the boot clamp retaining tabs (arrow) up with a small screwdriver, open the clamps and slide them off the boot

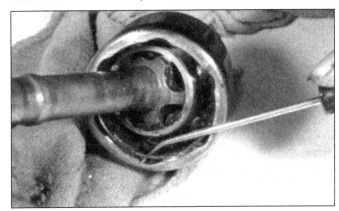

11.4 Pry the wire ring ball retainer out of the outer race

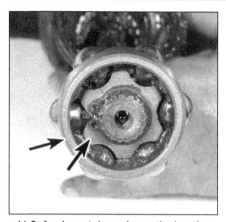

11.6 Apply match marks on the bearing (arrows) to identify which side faces out during reassembly

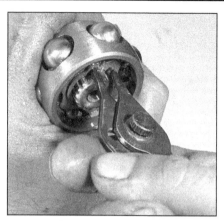

11.7 Remove the snap-ring from the end of the axle

11.10 Wrap the splined area of the axle with tape to prevent damage to the boot when installing it

5   Pull the outer race off the inner bearing assembly.
6   Make match marks on the inner and outer portions of the bearing to identify which side faces out on assembly **(see illustration)**.
7   Remove the snap-ring from the groove in the axleshaft with a pair of snap-ring pliers **(see illustration)**.
8   Slide the inner bearing assembly off the axleshaft.

Inspection
9   Clean the components with solvent to remove all traces of grease. Inspect the cage, balls and races for pitting, score marks, cracks and other signs of wear and damage. Shiny, polished spots are normal and will not adversely affect CV joint performance.

Reassembly
Refer to illustrations 11.10, 11.16, 11.17a and 11.17b
10   Wrap the axleshaft splines with tape to avoid damaging the boot. Slide the small boot clamp and boot onto the axleshaft, then remove the tape **(see illustration)**.
11   Install the inner bearing assembly on the axleshaft with the previously made matchmarks facing out.
12   Install the snap-ring in the groove. Make sure it's completely seated by pushing on the inner bearing assembly.
13   Fill the outer race and boot with the specified type and quantity of CV joint grease (normally included with the new boot kit). Pack the inner bearing assembly with grease, by hand, until grease is worked completely into the assembly.
14   Slide the outer race down onto the inner race and install the wire ring retainer.
15   Wipe any excess grease from the axle boot groove on the outer race. Seat the small diameter of the boot in the recessed area on the

axleshaft. Push the other end of the boot onto the outer race.
16   Equalize the pressure in the boot by inserting a dull screwdriver between the boot and the outer race **(see illustration)**. Don't damage the boot with the tool.
17   Install the boot clamps **(see illustrations)**.
18   Install a new circlip on the inner CV joint stub axle.
19   Install the driveaxle as described in Section 10.

Outer CV joint and boot
Disassembly
Refer to illustration 11.21
20   Following Steps 1 through 8, remove the inner CV joint from the axleshaft.
21   Remove the outer CV joint boot clamps, using the technique described in Step 3. Slide the boot off the axleshaft **(see illustration)**.

Inspection
Refer to illustration 11.23
22   Thoroughly wash the inner and outer CV joints in clean solvent and blow them dry with compressed air, if available. **Note:** *Because the outer joint cannot be disassembled, it is difficult to wash away all the old grease and to rid the bearing of solvent once it's clean. But it is imperative that the job be done thoroughly, so take your time and do it right.*
23   Bend the outer CV joint housing at an angle to the driveaxle to expose
the bearings, inner race and cage **(see illustration)**. Inspect the bearing surfaces for signs of wear. If the bearings are damaged or worn, replace the driveaxle.

11.16 Equalize the pressure inside the boot by inserting a small, dull screwdriver between the boot and the outer race

11.17a To install the new clamps, bend the tang down and . . .

11.17b . . . fold the tabs over to hold it in place

11.21 Slide the boot away from the joint and off the axleshaft

11.23 After the old grease has been rinsed away and the solvent
has been blown out with compressed air, rotate the outer joint
housing through its full range of motion and inspect the bearing
surfaces for wear and damage - if any of the balls, the race
or the cage look damaged, replace the driveaxle
and outer joint assembly

Reassembly

24 Slide the new outer boot onto the driveaxle. It's a good idea to
wrap vinyl tape around the spline of shaft to prevent damage to the
boot (see illustration 11.10). When the boot is in position, add the
specified amount of grease (included in the boot replacement kit) to
the outer joint and the boot (pack the joint with as much grease as it
will hold and put the rest into the boot). Slide the boot on the rest of the
way and install the new clamps (see illustrations 11.17a and 11.17b).
25 Proceed to clean and install the inner CV joint and boot by
following Steps 9 through 18, then install the driveaxle as outlined in
Section 10.

12 Front differential and driveshaft (4WD models) - removal, inspection and installation

Removal

Refer to illustrations 12.3a, 12.3b, 12.4 and 12.6

1   Refer to Section 10 and remove both front driveaxles.
2   Remove the front fender (see Chapter 7).
3   Unbolt the driveshaft cover and lift it off (see illustrations).
4   Remove the forward mounting bolt (see illustration 12.3a), then
unbolt the mounting bolt bracket from the frame (see illustration).

12.3a Remove the bolts at the front of the driveshaft cover
(there's one on each side) . . .

A) Driveshaft cover bolt   C) Upper mounting bolt and nut
B) Front mounting bolt and nut   D) Rear mounting bolt and nut

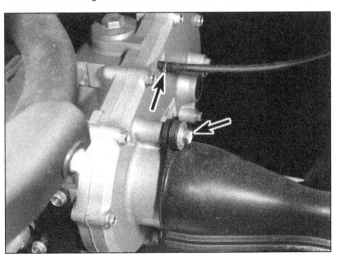

12.3b . . . and a bolt with rubber bushing at the rear (arrow) -
disconnect the breather hose (arrow) if you're
removing the transfer case

12.4 Remove the bolts (arrows) and the front bracket

12.6 The end of the driveshaft marked ENG goes
in the transfer case

12.8 Check the driveshaft oil seal on each side of the differential
(arrow) for wear or signs of leakage

5   Remove the nut from the upper mounting bolt and remove the bolt and collar **(see illustration 12.3a)**.
6   Remove the rear mounting bolt and nut. Pull the differential forward to disengage the driveshaft from the transfer case **(see illustration)**, then lift out the differential and driveshaft.

Inspection

Refer to illustration 12.8
7   Check the driveshaft for bending and for worn or damaged splines. Replace it if these conditions are found.
8   Check the oil seals at the driveaxle holes and the driveshaft hole for signs of leakage. Look into the driveaxle holes and check for obvious signs of wear and for damage such as broken gear teeth **(see illustration)**. Turn the pinion (where the driveshaft enters the differential) by hand (slip the driveshaft back on and use it as a handle if necessary).
9   Differential overhaul is a complicated procedure that requires several special tools, for which there are no readily available substitutes. If there's visible wear or damage, or if the differential's rotation is rough or noisy, take it to a Honda dealer for disassembly and further inspection.

Installation

10   Lubricate the lips of the driveaxle seals and the driveshaft seal, as

13.1 Remove the bolts (arrows) and lift the cover off

well as the splines of the driveshaft, with moly-based multi-purpose grease.
11   Place the differential in the frame, slightly forward of its installed position.
12   Slip the driveshaft into the differential. The end of the driveshaft marked ENG goes toward the transfer case **(see illustration 12.6)**.
13   Slide the differential back and slip the driveshaft into the transfer case.
14   Install the front mounting bracket **(see illustration 12.4)** and tighten its bolts to the torque listed in this Chapter's Specifications.
15   Install the front mounting bolt and nut **(see illustration 12.3a)** and tighten them to the torque listed in this Chapter's Specifications.
16   Install the upper mounting bolt and collar. Install the nut and tighten it to the torque listed in this Chapter's Specifications.
17   Install the rear mounting nut and bolt and tighten them to the torque listed in this Chapter's Specifications.
18   Fill the differential with the recommended type and amount of oil (see Chapter 1).
19   The remainder of installation is the reverse of the removal steps.

---

13 Front drive side shaft (4WD models) removal, inspection and installation

---

Removal

Refer to illustrations 13.1, 13.2a, 13.2b and 13.3
1   Unbolt the front drive side shaft cover and lift it off **(see illustration)**.
2   Remove the snap-ring from the groove in each end of the shaft **(see illustration)**.
Slide the snap-rings and washers along the shaft **(see illustration)**.
3   Tap the joint at each end of the shaft away from the transfer case (front end of shaft) and engine output shaft (rear end of shaft) with a hammer and punch **(see illustration)**. This shortens the shaft so it can be removed from the transfer case and output shaft splines.

Inspection

Refer to illustration 13.4
4   Check the splines in the joints for wear and damage **(see illustration)**.
5   Check the shaft for bending or other damage and replace it as necessary.
6   Check the cover for dents or bending and repair or replace it as necessary.

13.2a Remove the snap-ring at each end of the shaft . . .

13.2b . . . slide back the snap-ring and collar . . .

13.3 . . . and tap the joints out of the transfer case and output gear with a hammer and punch

Installation

7   Lubricate the shaft splines with multi-purpose grease.
8   Position the front drive side shaft between the output shaft and transfer case. Slide it back into the output shaft, then install the rear collar and snap-ring. Make sure the snap-ring seats securely in its groove.
9   Extend the front drive side shaft into the transfer case. If necessary, roll the vehicle or turn the front wheels to align the transfer case and shaft splines. Install the collar and front snap-ring, making sure the snap-ring seats securely.
10   Install the cover and tighten its bolts to the torque listed in this Chapter's Specifications.

14   Transfer case (4WD models) - removal and installation

Removal

Refer to illustrations 14.4a, 14.4b and 14.5

1   Remove the front fender (see Chapter 7) and drain the transfer case oil (see Chapter 1).
2   Remove the front driveshaft cover and disconnect the transfer case breather hose **(see illustration 12.3b)**.

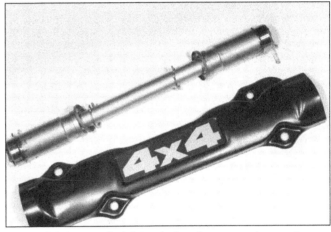

13.4 Inspect the joints, shaft, collars, snap-rings and cover for wear or damage

3   Remove the front drive side shaft (see Section 13).
4   Remove the mounting bolts from both sides of the transfer case **(see illustrations)**.

14.4a Remove the bolts from the left side; the upper and lower bolts (A) are a special type with spiral shanks; the center bolt (B) is a standard type

14.4b Remove the bracket from the right side

14.5 Push the transfer case and driveshaft forward, then pull the transfer case back and away from the vehicle

15.1a Pry the circlip out of its groove . . .

15.1b . . . and pull the driveshaft joint off the shaft

5   Push the transfer case forward, pushing the driveshaft into the front differential **(see illustration)**. Pull the transfer case back and sideways, separate it from the driveshaft and take it out of the frame.

Installation

6   Installation is the reverse of the removal steps, with the following additions:

　a) *Lubricate the driveshaft splines with multi-purpose grease. Make sure the end of the driveshaft marked ENG faces the transfer case* **(see illustration 12.6)**.

　b) *Tighten the mounting bolts to the torque listed in this Chapter's Specifications.*

　c) *Fill the transfer case with the recommended amount and type of oil* (see Chapter 1).

---

15   Transfer case (4WD models) - disassembly, inspection and reassembly

---

Disassembly

Refer to illustrations 15.1a, 15.1b, 15.2a, 15.2b and 15.3

1   Pry the circlip out of its groove and pull the driveshaft joint off its shaft **(see illustrations)**.

2   Remove the cover bolts and the oil drain bolt, then lift off the cover **(see illustrations)**.

3   Lift out the drive and driven gears and their shafts. Remove the gasket and dowels and clean all old gasket material from the cover mating surfaces **(see illustration)**.

Inspection

Refer to illustrations 15.4, 15.5, 15.7a and 15.7b

4   Check the oil seals for signs of leakage or wear **(see illustration)**. If there's any doubt about the condition of the seals, pry them out and tap in new ones with a seal driver or a socket the same diameter as the seal.

5   Spin the bearings in the case with a finger **(see illustration 15.4 and the accompanying illustration)**. If they're rough, loose or noisy, replace them.

6   To replace the driven gear bearing in the front cover, pry out the oil seal. Drive the bearing out with a bearing driver, then drive a new one in with a bearing driver or a socket that bears against the bearing outer race **(see illustration 15.4)**.

7   To replace the drive gear bearing in the front cover or the bearings in the rear cover, remove them with a blind hole puller and slide hammer **(see illustrations)**. Note that there's a washer behind the drive gear bearing in the front cover. Drive the new bearing(s) in with a bearing driver or a socket that bears against the bearing outer race.

8   Inspect the rubber plug in the front cover for signs of leakage and replace it as necessary.

Assembly

9   Assembly is the reverse of disassembly, with the following additions:

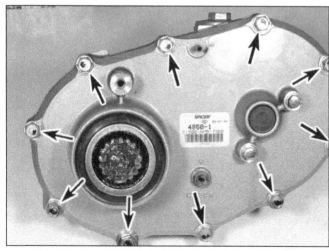

15.2a Remove the drain bolt and cover bolts (arrows) . . .

15.2b . . . and take the cover off

15.3 Note the locations of the dowels (arrows)

15.4 Check the seal (A) and bearing (B) for wear or damage

a) Lubricate the components with the recommended oil (see Chapter 1).
b) Use a new cover gasket and be sure the dowels are in position.
c) Tighten the cover bolts evenly, in a criss-cross pattern, to the torque listed in this Chapter's Specifications.
d) Use a new sealing washer on the drain bolt and tighten it to the torque listed in the Chapter 1 Specifications.
e) Be sure the circlip for the driveshaft joint is lodged securely in its groove.
f) After installation, fill the transfer case with the recommended type and quantity of oil (see Chapter 1).

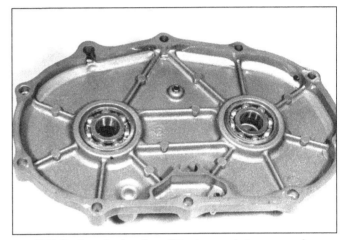

15.5 Replace the bearings if they're rough, loose or noisy when the inner race is rotated

---

16 Rear axle - removal, inspection and installation

Removal

Refer to illustrations 16.3, 16.4, 16.5, 16.6, 16.8a, 16.8b, 16.8c and 16.9

1   Securely block the front wheels so the vehicle won't roll. Loosen the rear wheel nuts with the vehicle on the ground. Jack up the rear end and support it securely, positioning the jackstands so they won't obstruct removal of the axle. The supports must be secure enough so the vehicle won't be knocked off of them while the very tight axle nuts are loosened. Remove the rear wheels.
2   Remove the rear wheel hubs (see Chapter 6).
3   The axle is secured on both sides by pairs of nuts tightened against each other. Loosening the nuts requires a pair of special wrenches **(see illustration)**. Equivalent wrenches are available from aftermarket sources. Honda tool numbers are:

   1988 through 1994 US: 07916-958020A and 07916-958010A
   1988 through 1994 except US: 07916-9580200 and 07916-9580400
   1995: 07916-958020B and 07916-958010B

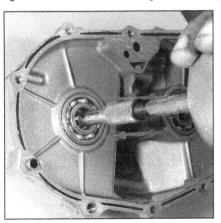

15.7a Remove the bearings with a puller and slide hammer . . .

15.7b . . . and drive the new ones in with a bearing driver or socket that bears against the bearing outer race

16.3 These special tools are used to remove the rear axle nuts

16.4 Place the holder portion of the tool on the inner nut . . .

16.5 . . . and loosen the outer nut with the other portion of the tool and a ratchet or breaker bar

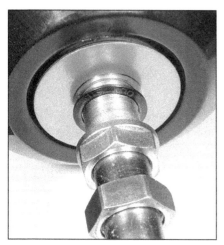

16.6 Unscrew the nuts and remove the washer; its OUT SIDE mark faces away from the brake drum cover on installation

16.8a Use the special tools to loosen the nuts on the left (final drive unit) side of the axle . . .

16.8b . . . remove the nuts, lockwasher (note the OUT SIDE mark) and washer (1998 and later models) . . .

4   Place the wrench over the inner nut on the brake drum side of the axle **(see illustration)**.

5   Place the wrench attachment on the outer nut and install a breaker bar or long ratchet handle in the square hole **(see illustration)**. Hold the wrench the turn the attachment counterclockwise to loosen the outer nut. **Note:** *If the wrench is hard to hold (the nuts are very tight), place it against the floor of the work area or on a solid support such as a large block of wood.*

6   Once the nuts are loose, unscrew them from the axle and remove the small and large washers **(see illustration)**.

7   Remove the brake drum cover, brake drum and brake panel (see Chapter 6).

8   On the other side of the axle, loosen the remaining nuts with the same tools **(see illustration)**. Remove the nuts, lockwasher, washer (1998 and later models) and O-ring **(see illustration)**.

9   Clean any foreign material from the left side of the axle (the opposite end from the brake panel) so it won't be pulled into the final drive unit during removal. Tap on the left end of the axle with a soft faced hammer to free it, then pull it out of the axle housing **(see illustration)**.

Inspection

Refer to illustration 16.11

10   Check the axle for obvious damage, such as step wear of the splines or bending, and replace it as necessary.

11   Install the wheel hubs on the axle. Place the axle in V-blocks and

16.8c . . . and remove the O-ring with a pointed tool

set up a pair of dial indicators contacting the bearing inner races **(see illustration)**. Rotate the axle and compare runout to the value listed in this Chapter's Specifications. If runout is excessive, replace the axle.

Installation

12   Lubricate the axle splines and seals with multi-purpose grease.

16.9 Pull the axle out from the right side; note the width of the smooth area just inside the splines (arrow) - on 1988 through 1992 models, the space on the left side of the axle is narrower; on 1993 and later models, the space is the same width on both sides of the axle

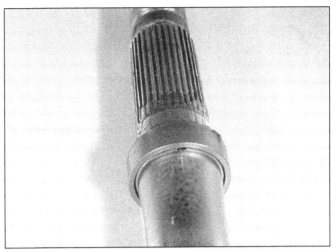

16.11 Check the bearing races for wear or damage

17.2a The axle housing is secured to the final drive by four bolts

17.2b Remove the bolts and pull the axle housing off the final drive unit . . .

16   Install the rear brake panel, shoes and drum (see Chapter 6).
17   Install the washer (1998 and later models) and lockwasher on the right side of the axle with its OUT SIDE mark facing out (see illustration 16.8b). Install the inner nut and tighten it to the torque listed in this Chapter's Specifications, using the special tools described above.
18   Prevent the inner nut from turning with the holder portion of the special tool. Apply non-permanent thread locking agent to the threads of the outer nut and tighten it against the inner nut to the torque listed in this Chapter's Specifications.
19   Repeat Steps 17 and 18 above to install the axle nuts on the left side of the vehicle.
20   The remainder of installation is the reverse of the removal steps.
21   Check oil level in the final drive unit and add oil as necessary (see Chapter 1).

17.2c . . . and remove the O-ring

17 Axle housing - removal and installation

Removal

Refer to illustrations 17.2a, 17.2b and 17.2c

1   Refer to Section 16 and remove the rear axle. Remove the trailer hitch (if equipped) (see Chapter 7).
2   Remove the axle housing bolts and pull the axle housing off the final drive unit (see illustrations). Remove the axle housing O-ring from the final drive unit (see illustration).

13   Install the axle from the right side of the vehicle. On 1988 through 1992 models, the narrow space inboard of the splines goes on the left side of the vehicle (see illustration 16.9).
14   Align the splines of the axle with those of the final drive unit.
15   Coat a new O-ring with oil and install it on the left side of the axle (see illustration 16.8b).

18.2 Remove the skid plate bolts (arrows) from the front and each side of the skid plate (right side bolt not shown)

18.3a Disconnect the breather tube and remove the four nuts (arrows) (one nut is hidden behind the swingarm); use new nuts on installation

18.3b Pull the final drive unit away from the swingarm and remove the O-ring (arrow)

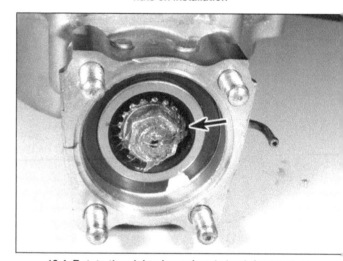

18.4 Rotate the pinion (arrow) and check for rough or noisy movement

Installation

3   Installation is the reverse of the removal steps, with the following additions:

  a) *Coat a new O-ring with oil and install it on the final drive unit.*
  b) *Tighten the axle housing bolts evenly, in a criss-cross pattern, to the torque listed in this Chapter's Specifications.*

18   Rear final drive unit - removal, inspection and installation

Removal

Refer to illustrations 18.2, 18.3a and 18.3b

1   Remove the axle shaft (see Section 16) and the axle housing (see Section 18).
2   Remove the skid plate from under the final drive unit **(see illustration)**.
3   Disconnect the breather tube and remove the nuts that secure the final drive unit to the swingarm **(see illustration)**. Pull the unit rearward to detach it and lift it away from the swingarm **(see illustration)**. Discard the nuts and use new ones on installation.

Inspection

Refer to illustration 18.4

4   Look into the axle holes and check for obvious signs of wear and for damage such as broken gear teeth. Also check the seals for signs of leakage. Turn the pinion **(see illustration)** by hand (slip the driveshaft back on and use it as a handle if necessary).
5   Differential overhaul is a complicated procedure that requires several special tools, for which there are no readily available substitutes. If there's visible wear or damage, or if the differential's rotation is rough or noisy, take it to a Honda dealer for disassembly and further inspection.

Installation

Refer to illustration 18.6

6   Make sure the driveshaft spring is in place in the driveshaft, then lubricate a new O-ring with oil and install it on the swingarm **(see illustration).**
7   Slide the final drive unit studs through the holes in the swingarm, align the driveshaft with the final drive pinion and push the final drive unit into position.
8   Install new final drive nuts and tighten them slightly.
9   Install the axle housing on the final drive unit and tighten its bolts to the torque listed in this Chapter's Specifications (see Section 17).

18.6 Be sure the spring is in place inside the driveshaft

20.7 Loosen the boot clamps so the boot can be detached from the output gear as the swingarm is removed

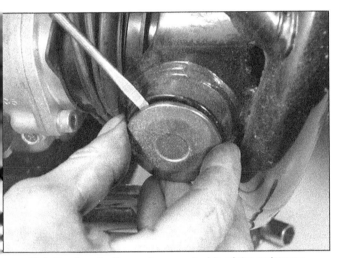

20.8 Pry the plastic cap from each side of the swingarm

20.9 Unscrew the left pivot bolt with a 17 mm Allen bolt bit

10   Tighten the final drive unit nuts to the torque listed in this Chapter's Specifications.
11   The remainder of installation is the reverse of the removal steps.

19 Swingarm bearings - check

1   Refer to Chapter 6 and remove the rear wheels, then refer to Section 4 and remove the rear shock absorber.
2   Grasp the rear of the swingarm with one hand and place your other hand at the junction of the swingarm and the frame. Try to move the rear of the swingarm from side-to-side. Any wear (play) in the bearings should be felt as movement between the swingarm and the frame at the front. The swingarm will actually be felt to move forward and backward at the front (not from side-to-side). If any play is noted, the bearings should be replaced with new ones (see Section 21).
3   Next, move the swingarm up and down through its full travel. It should move freely, without any binding or rough spots. If it does not move freely, refer to Section 21 for servicing procedures.

20 Swingarm - removal and installation

1   If the swingarm is being removed just for bearing replacement or driveshaft removal, the brake panel, final drive and rear axle need not

be removed from the swingarm. **Note:** *Loosening and tightening the pivot bolt locknut on the right side of the swingarm requires a special wrench for which there is no alternative. If you don't have the special Honda tool or an exact equivalent, the locknut must be unscrewed (and later tightened) by a Honda dealer.*

Removal

Refer to illustrations 20.7, 20.8, 20.9 and 20.10a through 20.10d
2   Raise the rear end of the vehicle off the ground with a jack. Support the vehicle securely so it can't be knocked over while it's jacked up.
3   Remove the rear wheels (see Chapter 7).
4   Refer to Section 4 and unbolt the lower end of the shock absorber from the swingarm.
5   Disconnect both rear brake cables (see Chapter 6).
6   If you're planning to remove the rear axle, final drive unit or brake panel, do it now (see Section 16, Section 18 or Chapter 6).
7   On the left side of the swingarm, loosen the clamps that secure the rubber boot **(see illustration)**.
8   Pry the plastic pivot cap from each side of the swingarm **(see illustration)**.
9   Unscrew the pivot bolt from the left side of the vehicle with a 17 mm Allen bolt bit **(see illustration)**. These are available from tool stores.
10   On the left side of the vehicle, prevent the pivot bolt from turning with the 17 mm Allen bolt bit and unscrew the pivot bolt locknut with

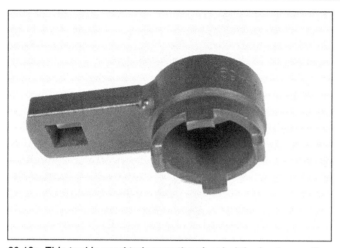

20.10a This tool is used to loosen the pivot bolt locknut as well as to tighten it to the correct torque

20.10b Insert the Allen bolt bit in the right pivot bolt . . .

20.10c . . . and place the special tool on the locknut; loosen the locknut with the special tool . . .

20.10d . . . then unscrew the locknut and pivot bolt

the special Honda tool **(see illustrations)**. If you're working on a 1988 through 1994 model, the tool number is 07908-4690001 (or KS-HBA-08-469 in the US). If you're working on a 1995 model, the tool number is 07980-469000A. Once the locknut is loose, unscrew the pivot bolt.

11   Pull the swingarm back and away from the vehicle, separating the driveshaft from the output gear as you pull.

12   Check the pivot bearings in the swingarm for dryness or deterioration (see Section 21). If they're in need of lubrication or replacement, refer to Section 21.

Installation

Refer to illustration 20.14

13   If the driveshaft was removed from the swingarm, install it. Lubricate the driveshaft splines with moly-base grease.

14   If the boot was removed from the swingarm, install it with one of its tabs pointing up **(see illustration)**. This aligns all three tabs with the heads of the Allen bolts on the output gear.

15   Lift the swingarm into position in the frame. Align the splines of the driveshaft with those of the output gear and align the pivot bolt holes in the swingarm with those in the frame.

16   Install the pivot bolt in the left side of the swingarm. Use the 17 mm Allen bolt bit to tighten it to the torque listed in this Chapter's Specifications.

17   Install the pivot bolt in the right side of the swingarm and tighten it to the torque listed in this Chapter's Specifications.

20.14 Install the boot with one of its tabs straight up

18   Raise and lower the swing arm several times, moving it through its full travel to seat the bearings and pivot bolts.

19   Retighten the pivot bolt in the right side of the swingarm to the torque listed in this Chapter's Specifications.

20   Hold the pivot bolt from turning with the 17 mm Allen bolt bit and use the locknut wrench to tighten the locknut to the torque listed in this

21.3 Pry the grease seals from the swingarm . . .

21.4 . . . to expose the bearings for inspection

22.2a Pull the driveshaft out of the swingarm . . .

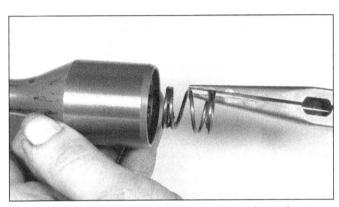

22.2b . . . and pull the spring out of the driveshaft

4   Rotate the center race of each ball bearing and check for roughness, looseness or play **(see illustration)**. If there's any doubt about bearing condition, replace the ball bearing, the sleeve and all three needle roller bearings as a set.

5   Inspect the needle roller bearings (use a flashlight if necessary). If there was play in the swingarm, or if there's any doubt about the condition of either bearing, replace both bearings as a set.

6   Bearing removal requires a blind hole puller; installation requires a drift of the exact same size as the bearings. If you don't have the correct tools, have the bearings replaced by a Honda dealer.

7   Pack the grease container inside each bearing with moly-based grease.

22   Rear driveshaft - removal, inspection and installation

Removal

Refer to illustrations 22.2a and 22.2b

1   Refer to Section 20 and remove the swingarm.

2   Pull the driveshaft out of the swingarm, then pull the spring out of the driveshaft **(see illustrations)**.

Inspection

Refer to illustrations 22.4 and 22.5

3   Check the shaft for bending or other visible damage such as step wear of the splines. If the shaft is bent, replace it. If the splines at the rear end of the shaft are worn, replace the shaft. If the splines at the front end of the shaft (the universal joint) are worn, replace the universal joint as described below.

4   Hold the driveshaft firmly in one hand and try to twist the universal joint **(see illustration)**. If there's play in the joint, replace it (but don't

22.4 If the universal joint is loose or clicks when it rotates, replace it

Chapter's Specifications. **Note:** *The reading shown on the torque wrench is lower than the actual locknut torque because the special tool increases the leverage of the torque wrench.*

21   Slip the boot over the output gear and tighten the clamp.

22   The remainder of installation is the reverse of the removal steps.

21   Swingarm bearings - replacement

Refer to illustrations 21.3 and 21.4

1   The swingarm pivot shaft rides on two ball bearings.

2   Remove the swingarm (see Section 20).

3   Pry the seal from each side of the swingarm **(see illustration)**.

confuse play in the joint with its normal motion).

5   To replace the joint, pry the circlip out of the groove (1988 through 1995 models only) and slide the joint off the shaft. Slide the new joint on and secure it with a new circlip, making sure it seats securely in its groove.

Installation

6   Installation is the reverse of the removal procedure.

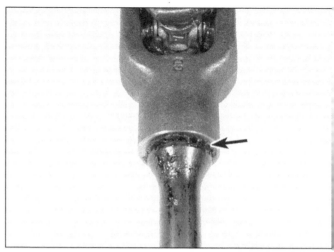

22.5 On 1988 through 1995 models, pry the circlip (arrow) out of the groove and use a new one on installation; slide the universal joint off the splines

Chapter 6
Brakes, wheels and tires

Contents

	Section
Brake fluid level check and fluid change	See Chapter 1
Brake hoses and lines - inspection and replacement	9
Brake lining wear check and system check	See Chapter 1
Brake pedal, lever and cables - removal and installation	13
Brake system bleeding	8
Front brake drums - inspection, waterproof seal replacement and 2WD bearing replacement	3
Front brake drums - removal and installation	2
Front brake master cylinder - removal, overhaul and installation	7
Front brake panel - removal, inspection and installation	6
Front brake shoes - removal, inspection and installation	4

	Section
Front wheel cylinders and 2WD adjusters - removal, overhaul and installation	5
General information	1
Rear brake drum - removal, inspection and installation	10
Rear brake panel - removal, inspection and installation	12
Rear brake shoes - removal, inspection and installation	11
Rear wheel hubs - removal and installation	16
Tires - general information	15
Tires and wheels - general check	See Chapter 1
Wheels - inspection, removal and installation	14

Specifications

Brakes

Brake fluid type	See Chapter 1
Brake lining minimum thickness	See Chapter 1
Brake pedal height	See Chapter 1
Front drum diameter	
2WD models	
Standard	130 mm (5.1 inches)
Wear limit	131 mm (5.2 inches)*
4WD models	
Standard	160 mm (6.299 inches)
Wear limit	161 mm (6.338 inches)*
Rear drum diameter	
Standard	160 mm (6.299 inches)
Wear limit	161 mm (6.338 inches)*
Front waterproof seal lubricant	
Type	NLGI no. 3
Amount	
2WD models	12 to 14 grams (0.4 to 0.5 oz)
4WD models	14 to 16 grams (0.5 to 0.6 oz)

* Refer to marks cast into the drum (they supersede information printed here)

Wheels and tires

Tire pressures	see Chapter 1
Tire tread depth	see Chapter 1

Torque specifications

Drum to hub bolts (4WD models)	10 Nm (84 in-lbs)
Front hub nuts	
2WD models..	60 to 80 Nm (43 to 58 ft-lbs)
4WD models..	80 to 100 Nm (58 to 72 ft-lbs)
Rear axle nuts..	see Chapter 5
Wheel cylinder bolts/nuts	
2WD models..	8 Nm (72 in-lbs)
4WD models	
6 mm...	8 Nm (72 in-lbs)
8 mm...	17 Nm (144 in-lbs)
Front brake adjuster bolts (2WD models)..	8 Nm (72 in-lbs)
Front brake hose union bolts..	35 Nm (25 ft-lbs)
Front brake hose joint nuts (2WD)	14 Nm (120 in-lbs)
Front brake pipe joint nuts..	14 Nm (120 in-lbs)
Master cylinder cover screws...	2 Nm (17 inch-pounds)
Master cylinder clamp bolts	12 Nm (108 in-lbs)
Brake panel bolts/nuts**	
Front brake..	30 Nm (22 ft-lbs)
Rear brake..	35 Nm (22 ft-lbs)
Rear brake panel drain bolt	25 Nm(18 ft-lbs)

**Discard the bolts or nuts and replace them with new ones each time they're removed.*

1 General information

The vehicles covered by this manual are equipped with hydraulic drum brakes on the front wheels and a single mechanical drum brake mounted on the rear axle inboard of the rear wheels. All brakes are sealed to keep water out. The front brakes are actuated by a lever on the right handlebar. The rear brakes have two means of actuation: a lever on the left handlebar and a pedal on the right side of the vehicle. The pedal and lever are connected to the rear brake assembly by cables.

All models are equipped with steel wheels, which require very little maintenance and allow tubeless tires to be used. **Warning:** *Brake components rarely require disassembly. Do not disassemble components unless absolutely necessary. If any hydraulic brake line connection in the system is loosened, the system must be bled of air. Do not use petroleum-based solvents when cleaning brake components. Use only clean brake fluid, brake system cleaner or isopropyl alcohol for cleaning. Use care when working with brake fluid as it can injure your eyes and it will damage painted surfaces and plastic parts.*

2 Front brake drums - removal and installation

Warning: *The dust created by the brake system may contain asbestos, which is harmful to your health (Honda hasn't used asbestos in brake parts for a number of years, but aftermarket parts may contain it). Never blow it out with compressed air and don't inhale any of it. An approved filtering mask should be worn when working on the brakes. Do not, under any circumstances, use petroleum-based solvents to clean brake parts. Use clean brake fluid, brake system cleaner or isopropyl alcohol only!*

Removal

Refer to illustrations 2.3 and 2.5

1   Loosen the front wheel nuts. Securely block the rear wheels so the vehicle can't roll. Raise the front end and support it securely on jackstands.
2   Refer to Section 14 and remove the front wheel.
3   Remove the cotter pin from the front hub nut **(see illustration)**. Have an assistant hold the brake on and remove the nut with a socket and breaker bar. If you're working on a 4WD model and you plan to

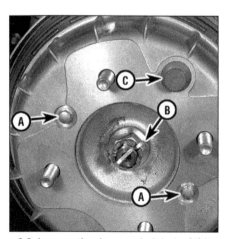

2.3 Loosen the drum-to-hub bolts (A) if you plan to separate the drum and hub; bend back the cotter pin (B) and remove the nut; the rubber plug (C) is for front brake adjustment

2.5 Replace the O-ring if the drum is separated from the hub (4WD models)

2.6a Pack grease into the space between the seal lips (the correct amount is listed in this Chapter's Specifications); be careful not to get any inside the drum

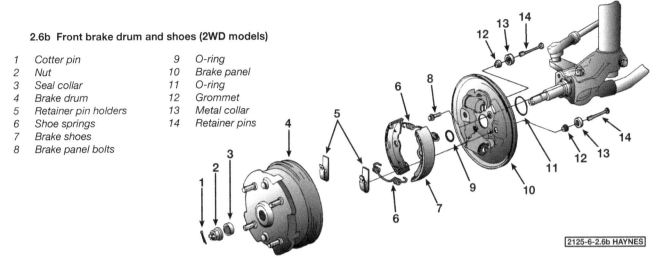

2.6b Front brake drum and shoes (2WD models)

1	Cotter pin	9	O-ring
2	Nut	10	Brake panel
3	Seal collar	11	O-ring
4	Brake drum	12	Grommet
5	Retainer pin holders	13	Metal collar
6	Shoe springs	14	Retainer pins
7	Brake shoes		
8	Brake panel bolts		

2125-6-2.6b HAYNES

inspect the drum-to-hub O-ring, loosen the drum-to-hub bolts at this point.

4   Pull the brake drum off.

5   If you're working on a 4WD model, remove the two drum-to-hub bolts and separate the drum from the hub **(see illustration 2.3)**. Inspect the hub O-ring **(see illustration)**. Since the O-ring's purpose is to keep water out of the brakes, replace it if its condition is in doubt.

Installation

Refer to illustrations 2.6a and 2.6b

6   Installation is the reverse of the removal steps, with the following additions:

a) *Pack the space between the drum seal lips with multi-purpose grease* **(see illustration)**. *Be sure not to get any grease on the inside of the drum; if you do, clean it off with a non-residue solvent such as brake system cleaner or lacquer thinner.*

b) *Tighten the drum-to-hub bolts on 4WD models to the torque listed in this Chapter's Specifications.*

c) *On 2WD models, be sure to insert the collar in the dust seal* **(see illustration)**.

d) *Tighten the hub nut to the torque listed in this Chapter's Specifications and install a new cotter pin. If necessary, tighten the nut to align the hole in the spindle with the slots in the nut.*

e) *Once the nut is tightened properly, bend the cotter pin to secure it.*

f) *Refer to Chapter 1 and adjust the brakes.*

---

3   Front brake drums - inspection, waterproof seal replacement and 2WD bearing replacement

---

Inspection

Refer to illustration 3.1

1   Check the brake drum for wear or damage. Measure the diameter at several points with a drum micrometer (or have this done by a Honda dealer). If the measurements are uneven (indicating that the drum is out-of-round) or if there are scratches deep enough to snag a fingernail, replace the drum. The drum must also be replaced if the diameter is greater than that cast inside the drum **(see illustration)**. Honda recommends against machining brake drums.

2   Check the waterproof seal on the edge of the brake drum for wear (caused by rubbing against the brake panel). To do this, measure the length of the seal lip from the point where it contacts the brake drum to the point where it contacts the brake panel **(see illustration 3.1)**. Also check for damage such as cuts and tears. If the seal is worn or damaged, replace it as described below.

3.1 The maximum diameter (A) is cast inside the brake drum; measure the length of the seal from the lip (B) to the point where it contacts the shoulder on the outer circumference of the brake drum

3   If you're working on a 2WD model, remove the collar from the dust seal in the center of the brake drum **(see illustration 2.6b)**. Check the seal for wear or damage. If its condition is in doubt, pry it out and drive in a new seal with a socket the same diameter as the seal. Turn the bearing in the brake drum with a finger. If its movement is rough, noisy or loose, replace it as described below.

Waterproof seal replacement

4   This procedure is somewhat complicated and must be done correctly to ensure a watertight seal between the brake drum and panel. It also requires a hydraulic press and a press plate bigger than the brake drum. For 4WD models, an additional steel plate 140 mm (5.5 inches) in diameter and at least 10 mm (0.4 inch) thick is required to prevent damage to the brake drum. If you don't have the necessary equipment, have this procedure done by a Honda dealer.

5   Carefully pry the waterproof seal off the edge of the brake drum.

6   It's important to press the new seal onto the drum just far enough, but not too far. To make sure this happens, you'll need to calculate the final clearance between the outer circumference of the drum and the seal, as well as between the inner circumference of the drum and the seal. These clearances will become smaller as the seal is pressed onto the drum, so calculating the clearances in advance will let you know when the seal has been pressed on far enough.

7   To determine what the final clearances should be, measure as follows.

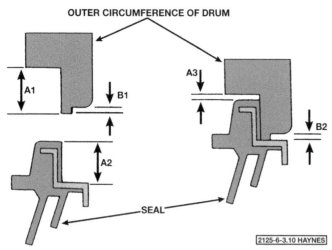

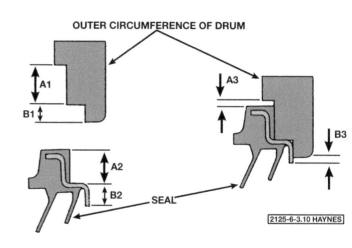

**3.8 On 2WD models, calculate the seal's installed clearances -
subtract measurement A2 from measurement A1 to get clearance
A3; clearance B2 is the same as measurement B1**

**3.10 On 4WD models, calculate the seal's installed clearances -
subtract measurement A2 from A1 to get clearance A3; subtract
measurement B2 from B1 to get clearance B3**

2WD models

Refer to illustration 3.8

8   Measure the depth of the recess in the brake drum and the seal surface that will contact it **(see illustration)**. Calculate the difference between these two measurements to get the final clearance for the outer circumference. Write this number down for later use.

9   Measure the depth of the thin portion of the brake drum that's closest to the brake panel when the drum is installed **(see illustration 3.8)**. This depth is the same as the final clearance for the inner circumference of the brake drum. Write the number down.

4WD models

Refer to illustration 3.10

10   Measure the depth of both recesses in the brake drum and the seal surfaces that will contact them **(see illustration)**. Calculate the difference between these measurements to get the final clearances for the seal as follows:

a) *Subtract measurement A2 from measurement A1 to get clearance A3.*
b) *Subtract measurement B2 from Measurement B1 to get clearance B3.*

11   Write the clearances down for later use.

All models

12   Dip the new seal in water to lubricate it (don't use any other type of lubricant). Place the new seal on the edge of the brake drum, then set the drum seal side down on a press plate.

13   If you're working on a 4WD model, place a steel plate about 5.5 mm (1.6 inch) in diameter and at least 10 mm (0.4 inch) thick on top of the brake drum so it won't collapse from the force of the press.

14   Slowly and carefully press the brake drum down into the seal, making sure not to press it too far. If the seal is damaged or is pressed on too far, remove it and start over with a new seal. Stop pressing when the measured clearances are equal to those written down in Steps 8 and 9 (2WD) or Step 11 (4WD).

2WD bearing replacement

15   The front wheel bearings on 2WD models are mounted in the brake drums.

16   Pry out the seal **(see illustration 2.6b)**.

17   Insert a soft metal drift into the hub from the inside. Tap gently against the inner bearing, on opposite sides of the bearing, to drive it from the drum. Then insert the drift from the other side and drive the outer bearing out in the same way.

18   Pack the new bearings with multi-purpose grease. Work the grease into the spaces between the bearing balls. Hold the outer race and rotate the bearing inner race as you pack it to distribute the grease.

19   Place the new outer bearing on the brake drum with its sealed side out. Position the drum seal side down on a workbench or similar surface. Support the center of the drum from below so the drum isn't resting on the waterproof seal. Tap the bearing into position with a bearing driver or socket the same diameter as the bearing outer race.

20   Turn the drum over and install the inner bearing in the same manner.

4   Front brake shoes - removal, inspection and installation

Removal

Refer to illustrations 4.2 and 4.3

1   Refer to Section 2 and remove the brake drum.

2   Rotate the retainer pins with pliers to align them with the slots in the pin holders, then remove the pin holders **(see illustration)**.

3   Pull the shoes apart against the force of the spring(s) and take them off the brake panel **(see illustration)**.

**4.2 Rotate the ends of the retainer pins to align with the slot in
the pin holder, then take the pin holders off**

4.3 Pull the shoes apart and take them off the brake panel

4.8 Inspect the rubber grommets on the retainer pins and replace them if they're damaged or deteriorated

4.11 The assembled brakes should look like this (4WD models)

Inspection

Refer to illustration 4.8

4   Check the linings for wear, damage and signs of contamination from road dirt or water. If the linings are visibly defective, replace them.

5   Measure the thickness of the lining material (just the lining material, not the metal backing) and compare with the value listed in the Chapter 1 Specifications. Replace the shoes if the material is worn to the minimum or less.

6   Check the ends of the shoes where they contact the wheel cylinders (4WD) or wheel cylinder and adjuster body (2WD). Replace the shoes if there's visible wear.

7   Pull back the wheel cylinder cups and check for fluid leakage. Slight moisture inside the cups is normal, but if fluid runs out, refer to Section 5 and replace the wheel cylinder(s).

8   Inspect the rubber seals on the retaining pins **(see illustration)**. Replace them if they're worn, hardened or deteriorated. The seals are meant to keep out water, not just dust, so replace them if there's any doubt about their condition.

Installation

Refer to illustration 4.11

9   Apply high temperature grease to the ends of the springs and shoes. Apply a thin smear of grease to each of the brake shoe contact

points on the brake panel.

10   Hook the springs to the shoes. If you're working on a 2WD model, the curved centers of the springs face out. If you're working on a 4WD model, the hooked ends of the springs face out.

11   Pull the shoes apart and position their ends in the wheel cylinder and adjuster (2WD) or wheel cylinders (4WD) **(see illustration)**. The flatter ends of the brake shoes fit into the wheel cylinder(s); the other ends fit into the adjuster(s).

12   Install the pin holders and pins. Compress the holders and turn the pins 90-degrees so the pins secure the holders.

13   The remainder of installation is the reverse of the removal steps.

---

5   Front wheel cylinders and 2WD adjusters - removal, overhaul and installation

---

Warning: *If a front wheel cylinder indicates the need for an overhaul (usually due to leaking fluid or sticky operation), both front wheel cylinders (2WD models) or all four front wheel cylinders (4WD models) should be overhauled and all old brake fluid flushed from the system. Also, the dust created by the brake system may contain asbestos, which is harmful to your health (Honda hasn't used asbestos in brake parts for a number of years, but aftermarket parts may contain it). Never*

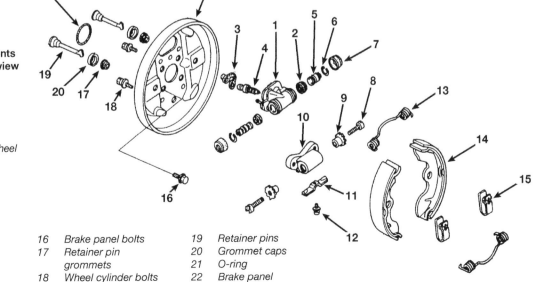

5.2 Front brake components (2WD models) - exploded view

1	Wheel cylinder body	
2	Piston cup seals	
3	Bleed valve cap	
4	Bleed valve	
5	Piston	
6	Piston clips (Nissin wheel cylinders)	
7	Boot	
8	Adjuster screw	
9	Adjuster wheel	
10	Adjuster body	
11	Lock spring	

12	Screw	16	Brake panel bolts	19	Retainer pins
13	Brake shoe springs	17	Retainer pin grommets	20	Grommet caps
14	Brake shoes	18	Wheel cylinder bolts	21	O-ring
15	Retainer pin holders			22	Brake panel

5.3a Unscrew the union bolt; use a new sealing washer on each side of the bolt on installation and position the neck of the hose between the stoppers (arrow)

5.3b Unscrew the brake pipe fitting (A) from the wheel cylinder with a flare nut wrench; remove the bolt and the nut (B)

5.3c Loosen the other end of the brake pipe and remove the nut and bolt

blow it out with compressed air and don't inhale any of it. An approved filtering mask should be worn when working on the brakes. Do not, under any circumstances, use petroleum-based solvents to clean brake parts. Use clean brake fluid, brake system cleaner or isopropyl alcohol only!

Removal

Refer to illustrations 5.2, 5.3a, 5.3b and 5.3c

1   Refer to Sections 2 and 4 and remove the brake drum and shoes.

2   2WD models use a single wheel cylinder between the upper ends of the shoes and an adjuster between the lower ends of the shoes **(see illustration)**. 4WD models use two wheel cylinders which incorporate adjuster mechanisms **(see illustration 4.11)**.

3   From the back side of the brake panel, detach the brake hose and metal line from the wheel cylinder(s). On 1988 through 1994 2WD models, hold the brake hose fitting with one wrench and unscrew the brake hose joint nut from the wheel cylinder with another wrench. On all other models, remove the union bolt and sealing washers and unscrew the brake pipe nuts **(see illustrations)**. Remove the wheel cylinder bolts (2WD) or nut and bolt (4WD). Work the wheel cylinder free of the sealant that secures it to the brake panel and take it off. If you're working on a 2WD model, unbolt the adjuster body and remove it from the brake panel.

Overhaul

Refer to illustrations 5.4a and 5.4b

4   Remove the boot(s) from the cylinder(s) **(see illustration)**. If you're working on a 2WD model equipped with Nissin wheel cylinders, remove the piston clips. If you're working on a 4WD model, remove the adjuster lock spring and pull the adjuster out of the other end of the cylinder **(see illustration)**.

5   Push the piston(s) out of the cylinder.

6   Check the piston and cylinder bore for wear, scratches and corrosion. If there's any doubt about their condition, replace the cylinder as an assembly. Even barely visible flaws can reduce braking performance.

7   The piston cups and boots are available separately and should be replaced whenever the wheel cylinders are overhauled. Work the cup(s) off the piston(s). Dip new ones in clean brake fluid and carefully install them without stretching or damaging them. The wide side of the piston cup faces into the cylinder bore.

8   Remove the screw from the adjuster nut (2WD models have two screws and two nuts). Check the adjuster components for wear or damage and replace as necessary.

9   Assembly is the reverse of the disassembly steps, with the

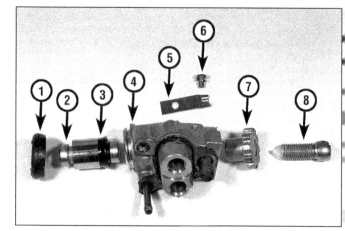

5.4a Wheel cylinder components (4WD models)

1	Boot	6	Screw
2	Piston	7	Adjuster wheel
3	Piston cup seal	8	Adjuster screw (left-
4	Wheel cylinder body		hand threads)
5	Lock spring		

following additions:

   a) Coat the cylinder bore with clean brake fluid. Install the piston with the wide side of the piston cup entering the bore first. Be sure not to turn back the lip of the cup.

   b) Make sure the piston clips on 2WD models with Nissin wheel cylinders are securely seated in the piston grooves.

   c) Lubricate the adjuster wheel(s) with silicone grease.

Installation

Refer to illustration 5.10

10   Installation is the reverse of the removal steps, with the following additions:

   a) Apply a thin coat of sealant to the wheel cylinder contact area on the brake panel **(see illustration)**. If you're working on a 2WD model, also apply sealant to the adjuster contact area.

   b) Be sure to install the wheel cylinders on the correct sides of the vehicle. Right wheel cylinders have the letter R cast into the cylinders body; left wheel cylinders are marked with an L.

   c) Tighten the wheel cylinder bolts (and nuts on 4WD models) to the torque listed in this Chapter's Specifications.

5.4b Remove the screw and take off the lock spring

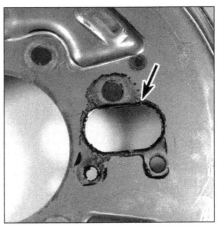

5.10 Apply sealant to the wheel cylinder contact areas (and the adjuster contact area on 2WD models)

6.2 Disconnect the breather hose

d) Use new sealing washers on the brake hose union bolt. Position the neck of the brake hose between the stoppers on the brake panel. Tighten the union bolt to the torque listed in this Chapter's Specifications and the metal brake line fittings securely.

e) Refer to Section 8 and bleed the brakes.

6   Front brake panel - removal, inspection and installation

Refer to illustrations 6.2 and 6.4

1   Refer to Section 2 and remove the brake drum. If you're planning to remove the brake shoes, refer to Section 4 (the brake panel can be removed with the shoes installed).

2   Disconnect the breather hose and remove the brake hose union bolt (see illustration). If you're planning to remove the wheel cylinders, this is a good time to loosen the bolts (and nuts on 4WD models), while the brake panel is securely bolted to the knuckle. The brake panel can be removed with the wheel cylinders installed.

3   Remove the brake panel mounting bolts. Lift the brake panel off the knuckle. Note: Discard the bolts and use new ones on installation. The bolt threads have a special dry waterproof coating.

4   Remove the O-ring from the knuckle and install a new one (see illustration).

5   Installation is the reverse of the removal steps, with the following additions:

a   Use new brake panel bolts and tighten them to the torque listed in this Chapter's Specifications.

b) Use new sealing washers on the brake hose union bolt (if equipped). Position the neck of the bolt between the stoppers on the brake panel and tighten the bolt to the torque listed in this Chapter's Specifications.

7   Front brake master cylinder - removal, overhaul and installation

1   If the master cylinder is leaking fluid, or if the lever doesn't produce a firm feel when the brake is applied, and bleeding the brakes does not help, master cylinder overhaul is recommended. Before disassembling the master cylinder, read through the entire procedure and make sure that you have the correct rebuild kit. Also, you will need some new, clean brake fluid of the recommended type, some clean rags and internal snap-ring pliers. Note: To prevent damage to the paint from spilled brake fluid, always cover the fuel tank when working on the master cylinder.

2   Caution: Disassembly, overhaul and reassembly of the brake master cylinder must be done in a spotlessly clean work area to avoid contamination and possible failure of the brake hydraulic system components.

Removal

Refer to illustrations 7.4 and 7.5

3   Loosen, but do not remove, the screws holding the reservoir cover in place.

4   Place rags beneath the master cylinder to protect the paint in

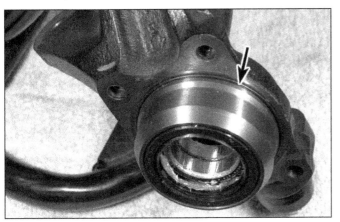

6.4 Install a new O-ring (arrow) on the knuckle

7.4 Remove the union bolt (arrow); on installation, use a new sealing washer on each side of the bolt

7.5 Remove the mounting bolts (arrows); the UP mark on the clamp must be upright when the clamp is installed

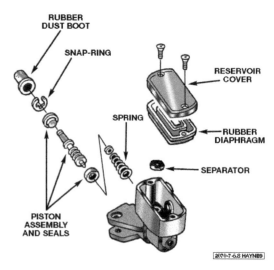

7.8 Master cylinder - exploded view

case of brake fluid spills. Remove the union bolt **(see illustration)** and separate the brake hose from the master cylinder. Wrap the end of the hose in a clean rag and suspend the hose in an upright position or bend it down carefully and place the open end in a clean container. The objective is to prevent excess loss of brake fluid, fluid spills and system contamination.

5   Remove the master cylinder mounting bolts **(see illustration)** and separate the master cylinder from the handlebar.

Overhaul

Refer to illustration 7.8

6   Remove the locknut from the underside of the lever pivot screw, then remove the screw.

7   Carefully remove the rubber dust boot from the end of the piston. Remove the separator from the bottom of the reservoir.

8   Using snap-ring pliers, remove the snap-ring **(see illustration)** and slide out the piston, the cup seals and the spring. Lay the parts out in the proper order to prevent confusion during reassembly.

9   Clean all of the parts with brake system cleaner (available at auto parts stores), isopropyl alcohol or clean brake fluid. **Warning:** *Do not, under any circumstances, use a petroleum-based solvent to clean brake parts.* If compressed air is available, use it to dry the parts thoroughly (make sure it's filtered and unlubricated). Check the master cylinder bore for corrosion, scratches, nicks and score marks. If damage is evident, the master cylinder must be replaced with a new one. If the master cylinder is in poor condition, then the wheel cylinders should be checked as well.

10   Remove the old cup seals and install the new ones. On 1988 through 1994 models, both cups are installed on the piston; on 1995 models, one cup seal is installed on the piston and the other cup seal is mounted on the end of the spring. Make sure the lips of the cup seals face away from the lever end of the piston. If a new piston is included in the rebuild kit, use it regardless of the condition of the old one.

11   Before reassembling the master cylinder, soak the piston and the rubber cup seals in clean brake fluid for ten or fifteen minutes. Lubricate the master cylinder bore with clean brake fluid, then carefully insert the piston and related parts in the reverse order of disassembly. Make sure the lips on the cup seals do not turn inside out when they are slipped into the bore.

12   Depress the piston, then install the snap-ring (make sure the snap-ring is properly seated in the groove with the sharp edge facing out). Install the rubber dust boot (make sure the lip is seated properly in the piston groove).

13   Install the brake lever and tighten the pivot bolt locknut.

Installation

14   Attach the master cylinder to the handlebar. Align the corner of the master cylinder's front mating surface with the punch mark on the handlebar.

15   Make sure the arrow and the word UP on the master cylinder clamp are pointing up, then tighten the bolts to the torque listed in this Chapter's Specifications. Tighten the top bolt fully, then tighten the lower bolt. **Caution:** *Don't try to close the gap at the lower bolt mating surface or the clamp may break.*

16   Connect the brake hose to the master cylinder, using new sealing washers. Tighten the union bolt to the torque listed in this Chapter's Specifications.

17   Refer to Section 8 and bleed the air from the system.

8   Brake system bleeding

Refer to illustration 8.5

1   Bleeding the brake system is simply the process of removing all the air bubbles from the brake fluid reservoir, the lines and the wheel cylinders. Bleeding is necessary whenever a brake system hydraulic connection is loosened, when a component or hose is replaced, or when the master cylinder or wheel cylinders are overhauled. Leaks in the system may also allow air to enter, but leaking brake fluid will reveal their presence and warn you of the need for repair.

2   To bleed the brake system, you will need some new, clean brake fluid of the recommended type (see Chapter 1), a length of clear vinyl or plastic tubing, a small container partially filled with clean brake fluid, some rags and a wrench to fit the brake wheel cylinder bleed valve.

3   Cover the fuel tank and other painted components to prevent damage in the event that brake fluid is spilled.

4   Remove the reservoir cap and slowly pump the brake lever a few times, until no air bubbles can be seen floating up from the holes at the bottom of the reservoir. Doing this bleeds the air from the master cylinder end of the line. Reinstall the reservoir cap.

5   Attach one end of the clear vinyl or plastic tubing to the wheel cylinder bleed valve and submerge the other end in the brake fluid in the container **(see illustration)**.

6   Check the fluid level in the reservoir. Do not allow the fluid level to drop below the lower mark during the bleeding process.

7   Carefully pump the brake lever three or four times and hold it while opening the bleed valve. When the valve is opened, brake fluid will flow out of the wheel cylinder into the clear tubing and the lever will move toward the handlebar.

8   Retighten the bleed valve, then release the brake lever gradually. Repeat the process until no air bubbles are visible in the brake fluid leaving the wheel cylinder and the lever is firm when applied. Remember to add fluid to the reservoir as the level drops. Use only new, clean brake fluid of the recommended type. Never re-use the fluid lost during bleeding.

8.5 Remove the rubber cap and connect a plastic or rubber hose to the bleed valve; open and close the valve with a wrench (this is the left front wheel)

9.2 Check the hoses for cracks; pay special attention to the points where they meet the metal fittings (arrows)

9   Repeat this procedure to the other wheel cylinder (4WD models) or to the wheel cylinder on the other wheel (2WD models). If you're working on a 4WD model, bleed the wheel cylinders on the other wheel only after both cylinders on the first wheel have been bled. Be sure to check the fluid level in the master cylinder reservoir frequently.

10   Replace the reservoir cap, wipe up any spilled brake fluid and check the entire system for leaks. **Note:** *If bleeding is difficult, it may be necessary to let the brake fluid in the system stabilize for a few hours (it may be aerated). Repeat the bleeding procedure when the tiny bubbles in the system have floated out.*

9   Brake hoses and lines - inspection and replacement

Inspection

Refer to illustration 9.2

1   Once a week, or if the vehicle is used less frequently, before every use, check the condition of the brake hoses.

2   Twist and flex the rubber hoses while looking for cracks, bulges and seeping fluid. Check extra carefully around the areas where the hoses connect with metal fittings, as these are common areas for hose failure **(see illustration)**.

Replacement

Refer to illustration 9.4

3   There are two brake hoses. One hose is attached to the master cylinder. On 2WD models, this hose connects to a metal pipe which in turn connects to the metal joint in the center of the lower hose. On 4WD models, the master cylinder hose runs all the way to the metal

joint. The metal joint is part of the lower hose assembly, consisting of the joint and two permanently attached lengths of hose which run to the wheel cylinders.

4   Cover the surrounding area with plenty of rags and unscrew the union bolts on either end of the hose. Detach the hose from any retainers that may be present and remove the hose **(see illustration)**.

5   Position the new hose, making sure it isn't twisted or otherwise strained, between the two components. Make sure the metal tube portion of the banjo fitting (if equipped) is located between the stoppers on the component it's connected to. Install the union bolts, using new sealing washers on both sides of the fittings, and tighten them to the torque listed in this Chapter's Specifications. If the hose is connected by a flare nut, hold it with one wrench and tighten the flare nut with another wrench.

6   Flush the old brake fluid from the system, refill the system with the recommended fluid (see Chapter 1) and bleed the air from the system (see Section 8). Check the operation of the brakes carefully before riding the vehicle.

10   Rear brake drum - removal, inspection and installation

Removal

Refer to illustrations 10.4a, 10.4b, 10.5 and 10.6

1   Refer to Section 14 and remove the right rear wheel.

2   Remove the right rear wheel hub (see Section 16).

3   Remove the skid plate from under the brake drum.

4   Remove the small, then the large washers from the axle shaft **(see illustrations)**.

9.4 Remove the retainers to release the hose (4WD shown)

10.4a Remove the small washer; the OUT SIDE mark (arrow) must face out on installation

10.4b Remove the large washer; replace the seal in the center of the drum cover if it shows signs of leakage

10.5 Remove the drum cover bolts and pull the cover off . . .

10.6 . . . then pull off the drum

10.7a Replace the small O-ring . . .

5   Remove the drum cover bolts and take the cover off **(see illustration)**.
6   Pull the drum off **(see illustration)**.

Inspection

Refer to illustrations 10.7a and 10.7b

7   Check the small and large O-rings for wear, deterioration or hardening **(see illustrations)**. Since the O-rings are intended to waterproof the rear brake, replace them if there's any doubt about their condition.
8   Inspect the seal in the brake drum cover for wear or signs of leakage **(see illustration 10.4b)**. If any defects are visible, pry the seal out and tap in a new one with a hammer and seal driver or a socket the same diameter as the seal.
9   Refer to Section 11 and look for leaks around the axle seal in the center of the brake panel. Replace the seal as described in Section 12 if there are any signs of leakage.
10   Refer to Section 3 for drum inspection details.

Installation

11   Installation is the reverse of the removal steps, with the following additions:

   a) *Be sure the large and small O-rings are in position* **(see illustrations 10.7a and 10.7b)**.
   b) *Coat the lip of the drum cover seal with grease.*
   c) *Be sure the OUT SIDE mark on the small washer faces out (away from the drum cover).*

11   Rear brake shoes - removal, inspection and installation

Removal

Refer to illustrations 11.2 and 11.3

1   Refer to Section 10 and remove the rear brake drum.
2   Bend back the cotter pin that secures the brake shoe washer and remove the washer **(see illustration)**.
3   Pull the brake shoes apart and fold them toward each other to release the spring tension **(see illustration)**. Remove the shoes and springs from the brake panel.

Inspection

Refer to illustration 11.5

4   Refer to Section 4 for shoe and lining inspection details.
5   Check the brake cam and anchor pin for wear or damage **(see illustration)**. If the brake cam shows wear or damage, refer to Section 12 and remove it from the brake panel. The anchor pin is cast into the brake panel, so the panel will have to be replaced if the pivot cups for the ends of the brake shoes are worn or damaged.

Installation

Refer to illustration 11.6

6   Installation is the reverse of the removal steps, with the following additions:

10.7b . . . and the large O-ring (arrow) if there's any doubt about their condition

11.2 Straighten the cotter pin and pull it out, then remove the washer

11.3 Pull the shoes apart and fold them into a V to release the spring tension, then take them off

11.5 Check the brake cam and the pivot cups in the anchor pin (arrows) for wear and damage

11.6 The assembled brake shoes should look like this

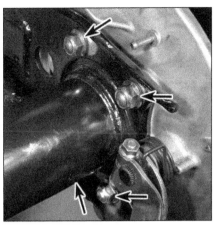

12.3 Remove and discard the brake panel nuts (arrows) (one nut is hidden behind the axle housing); use new nuts on installation

a) Apply a thin film of high-temperature brake grease to the brake cam and the pivot cups in the anchor pin, as well as to the shoe contact areas on the brake panel. Be sure not to get any grease on the brake drum or linings.
b) Place the shoes on the brake panel with their flatter ends against the brake cam (see illustration).
c) Secure the brake shoe retaining washer with a new cotter pin.

12 Rear brake panel - removal, inspection and installation

Removal

Refer to illustrations 12.3, 12.4 and 12.5

1   Refer to Sections 10 and 11 and remove the brake drum and shoes.
2   Refer to Section 13 and disconnect the brake cables from the brake panel lever.
3   Remove and discard the brake panel nuts (see illustration). Use new nuts on installation.
4   Disconnect the breather hose from the brake panel (see illustration).
5   Pull the brake panel away from the axle housing and take it off the axle shaft (see illustration).

12.4 Disconnect the breather hose . . .

Inspection

Refer to illustrations 12.6, 12.7, 12.8 and 12.9a through 12.9e

6   Spin the bearing in the brake panel with a finger (see illustration). If it's rough, loose or noisy, replace it as described below. Also check the seal for wear or damage and replace it if there's any doubt about its condition.
7   Pry out the seal with a removal tool or screwdriver (see illustration).

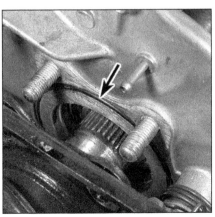

12.5 . . . and pull off the brake panel; replace the O-ring (arrow) if there's any doubt about its condition

12.6 If the bearing in the center of the brake panel is rough, loose or noisy, replace it

12.7 Pry out the seal with this special tool or a screwdriver

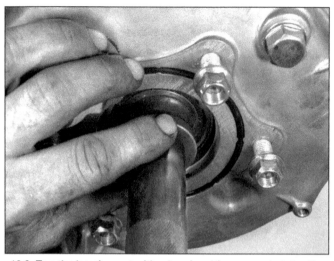

12.8 Tap the bearing out with a bearing driver or socket; tap the new bearing in with a bearing driver or socket that bears against the bearing outer race

12.9a Look for alignment marks on the cam and lever (arrows); make your own marks if necessary and unhook the spring (arrow)

12.9b Slide the wear indicator off the cam; the wide splines on cam and indicator (arrow) align with each other

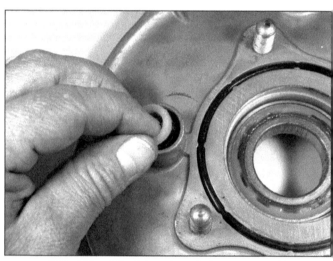

12.9c Lift out the felt seal; soak the new one with oil on installation

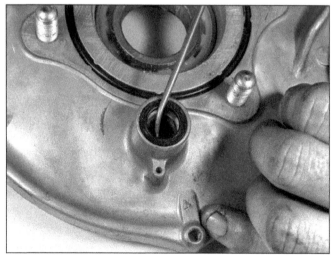

12.9d Pry out the rubber seal from behind the felt seal . . .

12.9e . . . and the rubber seal on the other side of the brake panel

12.13 If the brake panel drain bolt (arrow) is removed, install a new sealing washer

13.1 Remove the wing nuts and spacers, then slip the cables out of the slots (arrows)

8   Tap out the bearing from the O-ring side with a bearing driver or socket **(see illustration)**. Tap in the new bearing with a bearing driver or socket that bears against the bearing outer race. Install the bearing from the brake shoe side of the panel, with its black sealed side facing the brake shoe side. Position the new seal with its lip facing the bearing and tap it in with the same tool used to install the bearing.

9   If the brake cam is loose in its bore or if the seals show any signs of leakage, remove the brake cam from the panel **(see illustrations)**. Inspect the cam and its bore for wear and replace worn parts, then reverse the disassembly sequence to reinstall the brake cam.

10   Apply grease to the rubber seals and oil to the felt seal on assembly.

Installation

Refer to illustration 12.13

11   Install a new O-ring in the brake panel groove **(see illustration 12.5)**.

12   Position the brake panel on the axle housing, making sure the O-ring isn't dislodged. Install new brake panel nuts (don't re-use the old ones) and tighten them to the torque listed in this Chapter's Specifications.

13   If the brake panel drain bolt was removed, install it with a new sealing washer **(see illustration)** and tighten it to the torque listed in this Chapter's Specifications.

14   The remainder of installation is the reverse of the removal steps.

13   Brake pedal, lever and cables - removal and installation

Brake cables

Removal

Refer to illustration 13.1

1   Unscrew the cable adjusting nuts all the way off the end of the cable, then remove the cables from their slots in the brake panel **(see illustration)**.

2   Thread the adjusting hardware and wing nuts back onto the cables so they won't be lost.

Pedal cable

Refer to illustration 13.3

3   Pull the pedal cable housing back from the bracket on the frame near the right swing arm pivot, then slip the cable out through the slot **(see illustration)**.

4   Turn the cable 180-degrees forward and slip it out of the slot in the top of the brake pedal.

13.3 Pull the cable rearward and slip it out of the bracket (left arrow); then rotate it forward and slip it out of the pedal slot (right arrow)

Lever cable

5   Pull back the rubber boot from the left handlebar lever.

6   Refer to the brake cable adjustment procedure in Chapter 1 and back off the adjuster locknut at the handlebar lever, loosening the adjuster all the way.

7   Align the slots in the adjuster and its locknut with each other so they're facing directly away from the handlebar. Turn the cable out of the slots, then align it with the slot in the lever and slip it out.

Installation

8   Installation is the reverse of the removal steps, with the following additions:

   a) *Lubricate the cable ends with multi-purpose grease.*
   b) *Make sure the cables are secure in their slots and retainers.*
   c) *Adjust brake pedal and lever play as described in Chapter 1.*

Brake pedal

Removal

Refer to illustrations 13.10 and 13.12

9   Disconnect the pedal from the cable as described above.

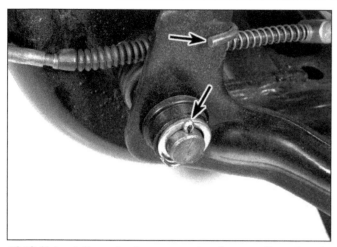

13.10 Unhook the spring (upper arrow) and remove the cotter pin and washer (lower arrow) . . .

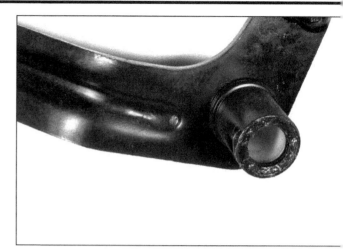

13.12 . . . then slip the pedal off the shaft and inspect the seal on each side of the pivot

10   Lift the pedal as far as possible and unhook the pedal spring **(see illustration)**.

11   Straighten the cotter pin, pull it out and remove the washer. Slide the pedal off the shaft **(see illustration 13.10)**.

12   Check the seal on each side of the pedal for wear or damage **(see illustration)**. If any defects are visible, pry the seal out and push new ones in.

Installation

13   Installation is the reverse of the removal steps, with the following additions:

    a) *Lubricate the pedal shaft and seal lips with multi-purpose grease.*
    b) *Refer to Chapter 1 and adjust brake pedal height.*

Brake lever

Removal

Refer to illustrations 13.16 and 13.17

14   Disconnect the cable from the brake lever as described above.

15   Refer to Chapter 2 and disconnect the reverse lock cable.

16   Look for a dot on the handlebar next to the parting line of the lever and clamp **(see illustration)**. This mark is used to position the lever correctly on the handlebar.

17   Remove the lever mounting screws and take it off the handlebar **(see illustration)**.

Installation

18   Installation is the reverse of the removal steps, with the following additions:

    a) *Make sure the handlebar and clamp parting line is aligned with the dot on the handlebar.*
    b) *Install the handlebar clamp with its dot up* **(see illustration 13.17)**. *Tighten the upper screw securely, then tighten the lower screw. Don't try to close the gap between the bottom of the clamp and the lever.*

---

14   Wheels - inspection, removal and installation

Inspection

1   Clean the wheels thoroughly to remove mud and dirt that may interfere with the inspection procedure or mask defects. Make a general check of the wheels and tires as described in Chapter 1.

2   The wheels should be visually inspected for cracks, flat spots on the rim and other damage. Since tubeless tires are involved, look very closely for dents in the area where the tire bead contacts the rim. Dents in this area may prevent complete sealing of the tire against the rim, which leads to deflation of the tire over a period of time.

3   If damage is evident, the wheel will have to be replaced with a new one. Never attempt to repair a damaged wheel.

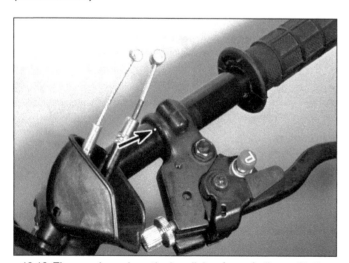

13.16 The punch mark on the handlebar (arrow) aligns with the seam in the lever bracket

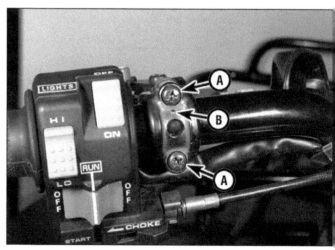

13.17 Remove the clamp screws (A); the dot on the clamp (B) is up when installed

14.7 The directional arrow on each tire must point in the forward rotating direction of the wheel; if it doesn't, the tire is mounted backward or the wheel is on the wrong side of the vehicle

14.8 Install the wheel nuts with their curved sides toward the wheel

Removal

4   Securely block the wheel at the opposite end of the vehicle from the wheel being removed, so it can't roll.
5   Loosen the lug nuts on the wheel being removed. Jack up one end of the vehicle and support it securely on jackstands.
6   Remove the lug nuts and pull the wheel off.

Installation

Refer to illustrations 14.7 and 14.8

7   Position the wheel on the studs. Make sure the directional arrow on the tire points in the forward rotating direction of the wheel **(see illustration)**.
8   Install the wheel nuts with their curved sides toward the wheel **(see illustration)**. This is necessary to locate the wheel accurately on the hub.
9   Snug the wheel nuts evenly in a criss-cross pattern.
10   Remove the jackstands, lower the vehicle and tighten the wheel nuts, again in a criss-cross pattern, to the torque listed in this Chapter's Specifications.

15   Tires - general information

1   Tubeless tires are used as standard equipment on this vehicle. Unlike motorcycle tires, they run at very low air pressures and are completely unsuited for use on pavement. Inflating ATV tires to excessive pressures will rupture them, making replacement of the tire necessary.
2   The force required to break the seal between the rim and the bead of the tire is substantial, much more than required for motorcycle tires, and is beyond the capabilities of an individual working with

normal tire irons or even a normal bead breaker. A special bead breaker is required for ATV tires; it produces a great deal of force and concentrates it in a relatively small area.
3   Also, repair of the punctured tire and replacement on the wheel rim requires special tools, skills and experience that the average do-it-yourselfer lacks.
4   For these reasons, if a puncture or flat occurs with an ATV tire, the wheel should be removed from the vehicle and taken to a dealer service department or a repair shop for repair or replacement of the tire. The illustrations on the following page can be used as a guide to tire replacement in an emergency, provided the necessary bead breaker is available.

16   Rear wheel hubs - removal and installation

Removal

1   Refer to Section 14 and remove the rear wheel(s).
2   Bend back the cotter pin and pull it out of the hub nut.
3   Unscrew the hub nut and remove the washer.
4   Pull the hub off the axle shaft.

Installation

5   Installation is the reverse of the removal steps, with the following additions:

 a) *Lubricate the axle shaft and hub splines with multi-purpose grease.*
 b) *Tighten the hub nut to the torque listed in this Chapter's Specifications. If necessary, tighten it an additional amount to align the cotter pin slots.*
 c) *Install a new cotter pin and bend it to secure the nut.*

Deflate the tire and remove the valve core. Release the bead on the side opposite the tire valve with an ATV bead breaker, following the manufacturer's instructions. Make sure you have the correct blades for the tire size (using the wrong size blade may damage the wheel, the tire or the blade). Lubricate the bead with water before removal (don't use soap or any type of lubricant).

TIRE CHANGING SEQUENCE

Turn the tire over and release the other bead.

If one side of the wheel has a smaller flange, remove and install the tire from that side. Use two tire levers to work the bead over the edge of the rim.

Before installing, ensure that tire is suitable for wheel. Take note of any sidewall markings such as direction of rotation arrows, then work the first bead over the rim flange.

Use tire levers to start the second bead over the rim flange.

Hold the bead while you work the last section of it over the rim flange. Install the valve core and inflate the tire, making sure not to overinflate it.

Chapter 7
Bodywork and frame

Contents

	Section		Section
Footpegs - removal and installation	6	Rear carrier and fender - removal and installation	5
Frame - general information, inspection and repair	8	Right side cover - removal and installation	4
Front carrier and fender - removal and installation	3	Seat - removal and installation	2
General information	1	Trailer hitch - removal and installation	7

1 General information

This Chapter covers the procedures necessary to remove and install the fenders and other body parts. Since many service and repair operations on these vehicles require removal of the fenders and/or other body parts, the procedures are grouped here and referred to from other Chapters.

In the case of damage to the fenders or other body parts, it is usually necessary to remove the broken component and replace it with a new (or used) one. The material that the fenders and other plastic body parts is composed of doesn't lend itself to conventional repair techniques. There are, however, some shops that specialize in "plastic welding", so it would be advantageous to check around first before throwing the damaged part away.

Note: *When attempting to remove any body panel, first study the panel closely, noting any fasteners and associated fittings, to be sure of returning everything to its correct place on installation. In most cases, the aid of an assistant will be required when removing panels, to help avoid damaging the paint. Once the visible fasteners have been removed, try to lift off the panel as described but DO NOT FORCE the panel - if it will not release, check that all fasteners have been removed and try again. Where a panel engages another by means of lugs and grommets, be careful not to break the lugs or to damage the bodywork. Remember that a few moments of patience at this stage will save you a lot of money in replacing broken panels!*

2 Seat - removal and installation

Refer to illustrations 2.1 and 2.2

1   Lift the seat latch **(see illustration)** and lift the back end of the seat.
2   Disengage the front end of the seat from the bracket **(see illustration)** and lift the seat off the vehicle.
3   Installation is the reverse of removal.

3 Front carrier and fender - removal and installation

Refer to illustrations 3.4 and 3.7

1   The front fender is a one-piece unit that spans the front of the vehicle and covers both front tires.
2   On 1988 through 1992 2WD models and 1988 4WD models, a separate lower fender panel is attached to the center unit on each side behind the front tire. On later 2WD models, these panels are permanently attached.
3   On later 4WD models, a separate outer/rear fender piece is attached to each side of the center unit.
4   Remove the carrier mounting bolts and lift the carrier off, taking care not to scratch the plastic fender **(see illustration)**.
5   Disconnect the headlight electrical connectors.

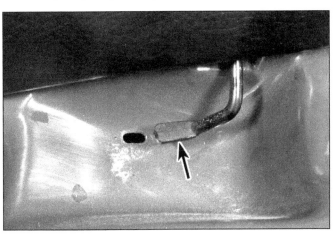

2.1 Lift the seat latch . . .

2.2 . . . and disengage the seat from the bracket

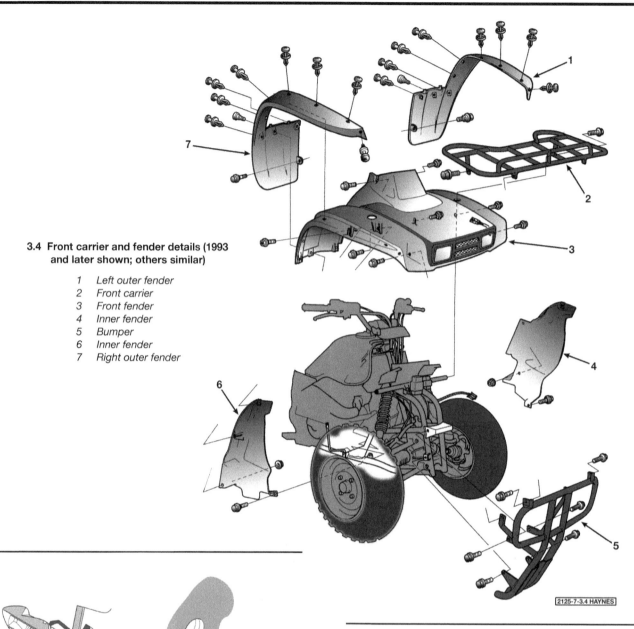

3.4 Front carrier and fender details (1993 and later shown; others similar)

1   *Left outer fender*
2   *Front carrier*
3   *Front fender*
4   *Inner fender*
5   *Bumper*
6   *Inner fender*
7   *Right outer fender*

3.7 Releasing the center pin type plastic insert trim clip

A) Press in center pin to unlock clip

B) Extract clip from panels

6   Remove the fender mounting bolts and lift the fender off the vehicle. Have an assistant support one side if necessary, so the fender can be lifted off without scratching it.

7   To remove the outer/lower fender panels on later models, release the trim clips. These clips are the "center pin" plastic insert type that are released by pressing in the clip's center pin with a pin punch or similar tool **(see illustration)**.

8   Installation is the reverse of removal.

4   Right side cover - removal and installation

Refer to illustration 4.1

1   Carefully pull the side cover from the vehicle. There are lugs that fit into rubber bushings on the frame, so apply a little extra force there (be careful not to break them off, though) **(see illustration)**.

2   Installation is the reverse of the removal procedure.

5   Rear carrier and fender - removal and installation

Refer to illustrations 5.2 and 5.3

1   Refer to *Battery - check* in Chapter 1 and remove the battery.

2   Disconnect the electrical connectors within the battery box **(see illustration)**. These will be pushed through the hole in the fender as it's removed.

3   Remove the rear carrier's mounting bolts and lift the carrier off the vehicle **(see illustration)**.

4   Open the toolbox cover. Remove the tool kit and remove the rear fender bolts inside the toolbox.

5   If you're working on a 1988 model, remove the metal fender stay from inside the wheel well on each side of the vehicle.

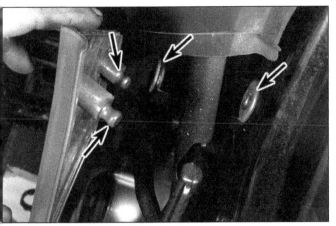

4.1 The right side cover is secured by lugs and grommets (arrows); be careful not to break the lugs on removal

5.2 Push the connectors through the hole in the fender on removal

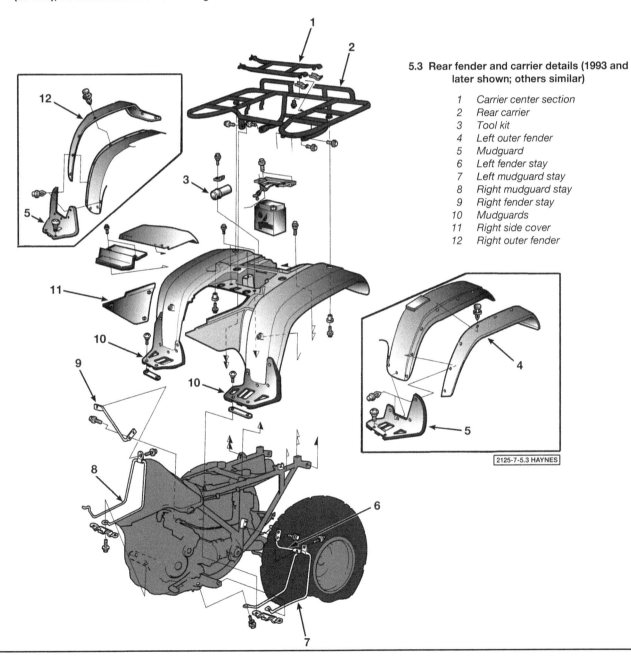

5.3 Rear fender and carrier details (1993 and later shown; others similar)

1   Carrier center section
2   Rear carrier
3   Tool kit
4   Left outer fender
5   Mudguard
6   Left fender stay
7   Left mudguard stay
8   Right mudguard stay
9   Right fender stay
10   Mudguards
11   Right side cover
12   Right outer fender

2125-7-5.3 HAYNES

6   If you're working on a 1989 or later model, remove the mudguard and mudguard stay from each side of the vehicle near the footpegs, then remove the rear fender stay from inside the wheel well on each side of the vehicle.

7   Carefully lift the rear fender off the vehicle, taking care not to scratch the plastic.

8   If necessary, remove the outer fenders from the center unit. On 1991 and later models, release the trim clips. These clips are the "center pin" plastic insert type that are released by pressing in the clip's center pin with a pin punch or similar tool **(see illustration 3.7)**.

9   Installation is the reverse of removal, with the following addition: If you're installing a new rear fender on a 1988 through 1992 model, you'll need to drill several 7.5 mm (0.3 inch) mounting holes. The hole locations are marked on the inside of the new fender. Drill at points marked A (US models) or C (Canadian models). Don't drill at points marked E. Note that the right side has more holes than the left side.

6.1 The footpeg mounting bolts are accessible from below (arrows); two of the four bolts are hidden behind the footpeg

6   Footpegs - removal and installation

Refer to illustration 6.1

1   To remove the footpeg, reach under the vehicle and remove its mounting bolts **(see illustration)**.

2   Install the footpeg, then install the mounting bolts from beneath the vehicle and tighten them securely.

7   Trailer hitch - removal and installation

Refer to illustration 7.1

1   To remove the trailer hitch, remove its mounting bolts, working from above and beneath the vehicle **(see illustration)**.

2   Install the trailer hitch, then install the mounting bolts and tighten them securely.

7.1 Two of the trailer hitch bolts (arrows) are accessible from above; the remaining three are accessible from below (the center bolt is hidden under the axle housing)

8   Frame - general information, inspection and repair

1   All models use a double-cradle frame made of cylindrical steel tubing.

2   The frame shouldn't require attention unless accident damage has occurred. In most cases, frame replacement is the only satisfactory remedy for such damage. A few frame specialists have the jigs and other equipment necessary for straightening the frame to the required standard of accuracy, but even then there is no simple way of assessing to what extent the frame may have been overstressed.

3   After the machine has accumulated a lot of miles, the frame should be examined closely for signs of cracking or splitting at the welded joints. Corrosion can also cause weakness at these joints. Loose engine mount bolts can cause ovaling or fracturing to the engine

mounting points. Minor damage can often be repaired by welding, depending on the nature and extent of the damage.

4   Remember that a frame that is out of alignment will cause handling problems. If misalignment is suspected as the result of an accident, it will be necessary to strip the machine completely so the frame can be thoroughly checked.

Chapter 8
Electrical system

Contents

	Section
Alternator stator coils and rotor- check and replacement	26
Battery - charging	4
Battery - check	See Chapter 1
Battery - inspection and maintenance	3
Brake light switches - check and replacement	10
Charging system - leakage and output test	25
Charging system testing - general information and precautions	24
Electrical troubleshooting	2
Fuses - check and replacement	5
General information	1
Handlebar switches - check	14
Handlebar switches - removal and installation	15
Headlight aim - check and adjustment	8
Headlight bulb - replacement	7
Ignition main (key) switch - check and replacement	13
Indicator bulbs - replacement	11

	Section
Kickstarter	See Chapter 2
Lighting system - check	6
Neutral and reverse switches - check and replacement	16
Oil temperature warning system - check and component replacement	12
Regulator/rectifier - check and replacement	27
Starter clutch - removal, inspection and installation	23
Starter diode - removal, check and installation	19
Starter motor - disassembly, inspection and reassembly	21
Starter motor - removal and installation	20
Starter reduction gears - removal, inspection, bearing replacement and installation	22
Starter relay - check and replacement	18
Starter switch - check and replacement	17
Tail light and brake light bulbs - replacement	9
Wiring diagrams	28

Specifications

Battery

Type	Maintenance free
Terminal voltage	13.0 to 13.2 volts

Bulbs

Headlights	25/25 watts
Tail light	5 watts
Indicator lights	1.7 watts
Brake light (if equipped)	21 watts

Charging system

Charging output voltage	13.5 to 15.5 volts at 5000 rpm
Charging output amperage	Zero to 5 amps at 5000 rpm
Current leakage limit	0.1 mA
Stator coil resistance	0.09 to 0.11 ohms at 20-degrees C (68-degrees F)

Starter motor

Brush length	
Standard	12.5 mm (0.49 inch)
Minimum	9.0 mm (0.35 inch)
Commutator diameter	Not specified
Circuit fuse ratings	15 amps

Oil temperature sensor resistance

At 25-degrees C (77-degrees F) .... 9.5 to 10.5 K-ohms
At 100-degrees C (212-degrees F) .... 0.95 to 1.05 K-ohms
At 170-degrees C (338-degrees F) .... 209 to 231 ohms

Torque specifications

Alternator cover bolts .... 10 Nm (84 in-lbs)
Alternator rotor bolt .... 110 Nm (80 ft-lbs)
Reverse/neutral rotor Allen bolt .... 12 Nm (108 in-lbs)*
Starter reduction gear cover bolts .... 10 Nm (84 in-lbs)
Starter clutch Torx bolts .... 16 Nm (144 in-lbs)*
Neutral and reverse switches .... 13 Nm (108 in-lbs)
Oil temperature sensor .... 18 Nm (156 in-lbs)

Apply non-permanent thread locking agent to the threads.

1   General information

The machines covered by this manual are equipped with a 12-volt electrical system. The components include a three-phase permanent magnet alternator and a regulator/rectifier unit. The regulator/rectifier unit maintains the charging system output within the specified range to prevent overcharging and converts the AC (alternating current) output of the alternator to DC (direct current) to power the lights and other components and to charge the battery.

An electric starter mounted to the engine case behind the cylinder is standard equipment. A kickstarter is also installed. The starting system includes the motor, the battery, the relay and the various wires and switches. If the engine kill switch and the main key switch are both in the On position, the circuit relay allows the starter motor to operate only if the transmission is in Neutral. **Note:** *Keep in mind that electrical parts, once purchased, can't be returned. To avoid unnecessary expense, make very sure the faulty component has been positively identified before buying a replacement part.*

2   Electrical troubleshooting

A typical electrical circuit consists of an electrical component, the switches, relays, etc. related to that component and the wiring and connectors that hook the component to both the battery and the frame. To aid in locating a problem in any electrical circuit, wiring diagrams are included at the end of this Chapter.

Before tackling any troublesome electrical circuit, first study the appropriate diagrams thoroughly to get a complete picture of what makes up that individual circuit. Trouble spots, for instance, can often be narrowed down by noting if other components related to that circuit are operating properly or not. If several components or circuits fail at one time, chances are the fault lies in the fuse or ground/earth connection, as several circuits often are routed through the same fuse and ground/earth connections.

Electrical problems often stem from simple causes, such as loose or corroded connections or a blown fuse. Prior to any electrical troubleshooting, always visually check the condition of the fuse, wires and connections in the problem circuit.

If testing instruments are going to be utilized, use the diagrams to plan where you will make the necessary connections in order to accurately pinpoint the trouble spot.

The basic tools needed for electrical troubleshooting include a test light or voltmeter, a continuity tester (which includes a bulb, battery and set of test leads) and a jumper wire, preferably with a circuit breaker incorporated, which can be used to bypass electrical components. Specific checks described later in this Chapter may also require an ammeter or ohmmeter.

Voltage checks should be performed if a circuit is not functioning properly. Connect one lead of a test light or voltmeter to either the negative battery terminal or a known good ground/earth. Connect the other lead to a connector in the circuit being tested, preferably nearest to the battery or fuse. If the bulb lights, voltage is reaching that point,

which means the part of the circuit between that connector and the battery is problem-free. Continue checking the remainder of the circuit in the same manner. When you reach a point where no voltage is present, the problem lies between there and the last good test point. Most of the time the problem is due to a loose connection. Since these vehicles are designed for off-road use, the problem may also be dirt, water or corrosion in a connector. Keep in mind that some circuits only receive voltage when the ignition key is in the On position.

One method of finding short circuits is to remove the fuse and connect a test light or voltmeter in its place to the fuse terminals. There should be no load in the circuit. Move the wiring harness from side-to-side while watching the test light. If the bulb lights, there is a short to ground/earth somewhere in that area, probably where insulation has rubbed off a wire. The same test can be performed on other components in the circuit, including the switch.

A ground/earth check should be done to see if a component is ground/earthed properly. Disconnect the battery and connect one lead of a self-powered test light (such as a continuity tester) to a known good ground/earth. Connect the other lead to the wire or ground/earth connection being tested. If the bulb lights, the ground/earth is good. If the bulb does not light, the ground/earth is not good.

A continuity check is performed to see if a circuit, section of circuit or individual component is capable of passing electricity through it. Disconnect the battery and connect one lead of a self-powered test light (such as a continuity tester) to one end of the circuit being tested and the other lead to the other end of the circuit. If the bulb lights, there is continuity, which means the circuit is passing electricity through it properly. Switches can be checked in the same way.

Remember that all electrical circuits are designed to conduct electricity from the battery, through the wires, switches, relays, etc. to the electrical component (light bulb, motor, etc.). From there it is directed to the frame (ground/earth) where it is passed back to the battery. Electrical problems are basically an interruption in the flow of electricity from the battery or back to it.

3   Battery - inspection and maintenance

1   Most battery damage is caused by heat, vibration, and/or low electrolyte levels, so keep the battery securely mounted, and make sure the charging system is functioning properly. The battery used on these vehicles is a maintenance free (sealed) type and therefore doesn't require the addition of water. However, the following checks should still be regularly performed. **Warning:** *Always disconnect the negative cable first and connect it last to prevent sparks which could the battery to explode.*

2   Refer to Chapter 1 for battery removal procedures.

3   Check the battery terminals and cables for tightness and corrosion. If corrosion is evident, disconnect the cables from the battery, disconnecting the negative (-) terminal first, and clean the terminals and cable ends with a wire brush or knife and emery cloth. Reconnect the cables, connecting the negative cable last, and apply a thin coat of petroleum jelly to the cables to slow further corrosion.

4   The battery case should be kept clean to prevent current leakage,

which can discharge the battery over a period of time (especially when it sits unused). Wash the outside of the case with a solution of baking soda and water. Do not get any baking soda solution in the battery cells. Rinse the battery thoroughly, then dry it.

5   Look for cracks in the case and replace the battery if any are found. If acid has been spilled on the frame or battery box, neutralize it with a baking soda and water solution, then touch up any damaged paint. Make sure the battery vent tube (if equipped) is directed away from the frame and is not kinked or pinched.

6   If acid has been spilled on the frame or battery box, neutralize it with the baking soda and water solution, dry it thoroughly, then touch up any damaged paint. Make sure the battery vent tube is directed away from the frame and is not kinked or pinched.

7   If the vehicle sits unused for long periods of time, disconnect the cables from the battery terminals. Refer to Section 4 and charge the battery approximately once every month.

8   The condition of the battery can be assessed by measuring the voltage present at the battery terminals; voltmeter positive probe to the battery positive terminal and the negative probe to the negative terminal. When fully charged there should be approximately 13 volts present. If the voltage falls below 12.3 volts the battery must be removed, disconnecting the negative cable first, and charged as described in Section 4.

9   Refer to Chapter 1 to install the battery.

4   Battery - charging

Refer to illustration 4.2

1   If the machine sits idle for extended periods or if the charging system malfunctions, the battery can be charged from an external source.

2   The battery should be charged at no more than the rate printed on the charging rate and time label fixed to the battery. Honda recommends a special battery tester and charger which are unlikely to be available to the vehicle owner. To measure the charging rate, connect an ammeter is series with a battery charger **(see illustration)**.

3   When charging the battery, always remove it from the machine.

4   Disconnect the battery cables (negative cable first), then connect a voltmeter between the battery terminals and measure the voltage.

5   If terminal voltage is within the range listed in this Chapter's Specifications, the battery is fully charged. If it's lower, recharge the battery.

6   A quick charge can be used in an emergency, provided the maximum charge rate and time printed on the battery are not exceeded (exceeding the maximum rate or time may buckle the battery plates, rendering it useless). A quick charge should always be followed as soon as possible by a charge at the standard rate and time.

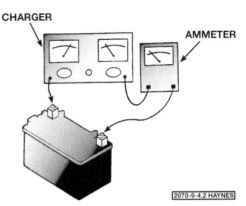

4.2 If the charger doesn't have an ammeter built in, connect one in series as shown; DO NOT connect the ammeter between the battery terminals or it will be ruined

7   Hook up the battery charger leads (positive lead to battery positive terminal, negative lead to battery negative terminal), then, and only then, plug in the battery charger. **Warning:** *The hydrogen gas escaping from a charging battery is explosive, so keep open flames and sparks well away from the area. Also, the electrolyte is extremely corrosive and will damage anything it comes in contact with.*

8   Allow the battery to charge for the specified time. If the battery overheats or gases excessively, the charging rate is too high. Either disconnect the charger or lower the charging rate to prevent damage to the battery.

9   After the specified time, unplug the charger first, then disconnect the leads from the battery.

10   If the recharged battery discharges rapidly when left disconnected, it's likely that an internal short caused by physical damage or sulfation has occurred. A new battery will be required. A sound battery will tend to lose its charge at about 1% per day.

11   When the battery is fully charged, unplug the charger first, then disconnect the leads from the battery. Wipe off the outside of the battery case and install the battery in the vehicle.

5   Fuses - check and replacement

Refer to illustrations 5.1a and 5.1b

1   There are two 15 amp fuses, located under the battery cover next to the battery **(see illustrations)**. The fuses are mounted in plastic holders, which are contained in rubber covers. Each rubber cover also contains a spare fuse.

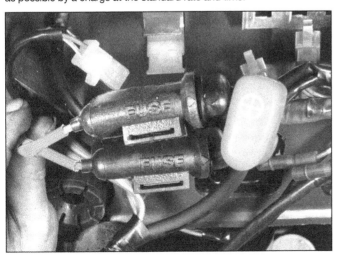

5.1a The fuse holders are located inside these covers near the battery . . .

5.1b . . . each cover contains a fuse holder (arrow) and a spare fuse

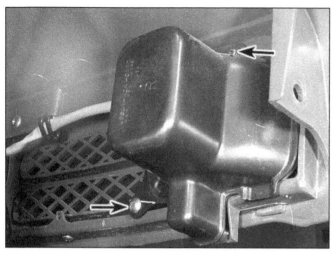

7.1 Remove the headlight cover screws (arrows), then pull back the rubber cover from the headlight case

7.3 Note the TOP mark on the rubber cover (left arrow), which must be up when the rubber cover is installed - push the bulb socket (right arrow) in and turn counterclockwise (anti-clockwise) to disengage the tabs, then pull it out . . .

2   The fuses can be removed and checked visually. Pull open the rubber cover, snap open the plastic holder and pull the fuse out. A blown fuse is easily identified by a break in the element.

3   If a fuse blows, be sure to check the wiring harnesses very carefully for evidence of a short circuit. Look for bare wires and chafed, melted or burned insulation. If a fuse is replaced before the cause is located, the new fuse will blow immediately.

4   Never, under any circumstances, use a higher rated fuse or bridge the fuse terminals, as damage to the electrical system could result.

5   Occasionally a fuse will blow or cause an open circuit for no obvious reason. Corrosion of the fuse ends and fuse holder terminals may occur and cause poor fuse contact. If this happens, remove the corrosion with a wire brush or emery paper, then spray the fuse end and terminals with electrical contact cleaner.

---

6   Lighting system - check

---

1   The battery provides power for operation of the headlights, tail light, brake light and instrument cluster lights. If none of the lights operate, always check battery voltage before proceeding. Low battery voltage indicates either a faulty battery, low battery electrolyte level or a defective charging system. Refer to Chapter 1 and Section 3 of this Chapter for battery checks and Sections 24 through 27 for charging system tests. Also, check the condition of the fuses and replace any blown fuses with new ones.

Headlights

2   If both of the headlight bulbs are out with the engine running (Maine and New Hampshire models) or with the lighting switch in the On position (all other models), check the main fuse with the key On (see Section 5).

3   If only one headlight is out, refer to Section 7 and unplug the electrical connector for the headlight bulb. Use a jumper wire to connect the bulb directly to the battery terminals as follows:

 a) Green wire terminal to battery negative terminal.
 b) White wire terminal to battery positive terminal.
 c) Blue wire terminal to battery positive terminal.

    If the light comes on, the problem lies in the wiring or one of the switches in the circuit. Refer to Sections 13 and 14 for the switch testing procedures, and also the wiring diagrams at the end of this Chapter. If either filament (high or low beam) of the bulb doesn't light, the bulb is burned out. Refer to Section 7 and replace it.

Tail light

4   If the tail light fails to work, check the bulb and the bulb terminals first, then check for battery voltage at the power wire in the tail light. If voltage is present, check the ground/earth circuit for an open or poor connection.

5   If no voltage is indicated, check the wiring between the tail light and the lighting switch, then check the switch.

Brake light

6   See Section 10 for the brake light circuit checking procedure.

Neutral indicator light

7   If the neutral light fails to operate when the transmission is in Neutral, check the main fuse and the bulb (see Section 11 for bulb removal procedures). If the bulb and fuse are in good condition, check for battery voltage at the wire attached to the neutral switch on the right side of the engine. If battery voltage is present, refer to Section 16 for the neutral switch check and replacement procedures.

8   If no voltage is indicated, check the wiring to the bulb, to the switch and between the switch and the bulb for open circuits and poor connections.

---

7   Headlight bulb - replacement

Refer to illustrations 7.1, 7.3 and 7.4

Warning: *If the headlight has just burned out, give it time to cool before changing the bulb to avoid burning your fingers.*

1   Reach inside the front fender, remove the headlight cover screws and take off the cover **(see illustration)**.

2   Pull the rubber dust cover off the headlight case.

3   Twist the bulb socket counterclockwise and remove it from the headlight case **(see illustration)**.

4   Pull the bulb out without touching the glass **(see illustration)**.

5   Installation is the reverse of the removal procedure, with the following additions:

 a) *Be sure not to touch the bulb with your fingers - oil from your skin will cause the bulb to overheat and fail prematurely. If you do touch the bulb, wipe it off with a clean rag dampened with rubbing alcohol.*

 b) *Align the tab on the metal bulb flange with the slot in the headlight case.*

 c) *Make sure the Top mark on the rubber cover is up **(see illustration 7.3)**.*

7.4 . . . and pull the bulb out of the headlight case without touching the glass

8.3 The headlight adjusting screw is located below each headlight (arrow)

8 Headlight aim - check and adjustment

Refer to illustration 8.3

1   An improperly adjusted headlight may cause problems for oncoming traffic or provide poor, unsafe illumination of the terrain ahead. Before adjusting the headlight, be sure to consult with local traffic laws and regulations. Honda doesn't provide specifications for headlight adjustment.

2   The headlight beam can be adjusted vertically. Before performing the adjustment, make sure the fuel tank is at least half full, and have an assistant sit on the seat.

3   Insert a Phillips screwdriver into the vertical adjuster screw **(see illustration)**, then turn the adjuster as necessary to raise or lower the beam.

9 Tail light and brake light bulbs - replacement

Tail light

Refer to illustrations 9.2a and 9.2b

1   Open the toolbox for access to the bulb socket.

2   Pull the bulb socket out of its housing, then pull the bulb out of the socket **(see illustrations)**.

3   Check the socket terminals for corrosion and clean them if necessary.

4   Make sure the rubber gasket is in place and in good condition, then push the bulb into the socket.

Brake light

5   If the vehicle is equipped with a brake light, remove the lens securing screws and take off the lens.

6   Press the bulb into its socket and turn it counterclockwise to remove.

7   Installation is the reverse of removal.

10 Brake light switches - check and replacement

Check

1   Before checking any electrical circuit, check the fuses (see Section 5).

2   Using a test light connected to a good ground/earth, check for voltage to the pink wire at the brake light switch. If there's no voltage present, check the pink wire between the switch and the ignition switch (see the wiring diagrams at the end of the book).

3   If voltage is available, touch the probe of the test light to the other terminal of the switch, then pull the brake lever or depress the brake pedal - if the test light doesn't light up, replace the switch.

4   If the test light does light, check the wiring between the switch and the brake lights (see the wiring diagrams at the end of the book).

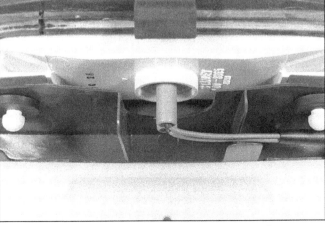

9.2a Pull the bulb socket out of the tail light housing . . .

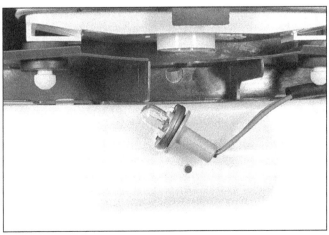

9.2b . . . and pull the bulb out of the socket

11.1 Pull the bulb socket out of the handlebar cover and pull the lens out of the socket . . .

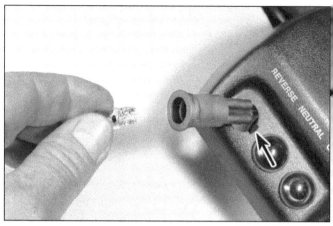

11.2 . . . then pull the bulb out; if the bulb is hard to reach, slip a piece of rubber tubing over it and pull on the tubing to remove the bulb; on later models, align the socket lugs with the cover grooves (arrow) on installation

Switch replacement

Brake lever switch

5   Unplug the electrical connectors from the switch.
6   Remove the mounting screw and detach the switch from the brake lever bracket/front master cylinder.
7   Installation is the reverse of the removal procedure. The brake lever switch isn't adjustable.

Brake pedal switch

8   Unplug the electrical connector in the switch harness.
9   Disconnect the spring from the brake pedal switch.
10   Hold the adjuster nut from turning and rotate the switch body all the way up until it clears the nut threads, then lift it out.
11   Install the switch by reversing the removal procedure.

---

11   Indicator bulbs - replacement

Refer to illustrations 11.1 and 11.2
1   To replace a bulb, pull the appropriate rubber socket out of the handlebar cover, then pull off the lens **(see illustration)**.
2   Pull the bulb out of the socket **(see illustration)**. **Note:** *To reach the bulbs, which are deep within the sockets, push a piece of rubber or soft plastic tubing down over the bulb, then pull the tubing and bulb out. Use the same tubing to install the new bulb.*
3   If the socket contacts are dirty or corroded, they should be scraped clean and sprayed with electrical contact cleaner before new bulbs are installed.
4   Carefully push the new bulb into position, using the same tubing that was used for removal, then push the socket into the handlebar cover. On 1994 and later models, align the lugs on the sockets with the slots in the cover **(see illustration 11.2)**.

---

12   Oil temperature warning system - check and component replacement

---

System check

1   These vehicles are equipped with an oil temperature warning system that turns on an indicator light on the handlebars when the oil overheats. The system consists of an oil temperature sensor mounted on the right side of the engine, an alarm unit mounted at the front of the vehicle and a warning indicator light in the handlebar cover. The indicator light should come on for a few seconds when the engine is first started, then turn off. If the oil temperature light on your vehicle comes on often due to operation in high temperatures, you can purchase a cooling fan kit from a Honda dealer. **Caution:** *If the light comes on while the engine is running, shut it off immediately and let it*

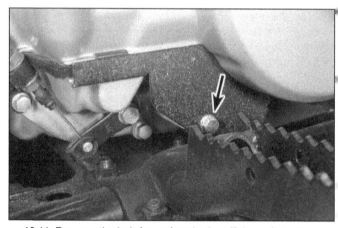

12.11 Remove the bolt (arrow) and take off the switch cover

cool. Continued operation with overheated oil can cause serious engine damage.
2   If the oil temperature warning light fails to operate properly, check the bulb and replace it if it's burned out.
3   If the bulb is good, check the electrical connections at the alarm unit. On 1988 through 1997 models, the alarm unit is a separate component with two connectors, one with four pins and one with two. It's mounted next to the ignition control module **(see illustration 5.2 in Chapter 4)**. On 1998 and later models, the alarm unit is built into the ignition control module. Its wires are included in the eight-pin ICM connector.
4   If the bulb and connectors are good, disconnect the two connectors from the alarm unit (1988 through 1997 models) or the single connector from the ignition control module (1998 and later models). Turn the ignition switch to the On position (but don't start the engine). Connect the negative probe of a voltmeter to the green wire's terminal in the harness side of the connector. Connect the voltmeter's positive probe to the brown-red wire's terminal, then to the black wire's terminal. The voltmeter should indicate battery voltage (approximately 13 volts) in both cases. If it doesn't, check the wiring for breaks or bad connections (see the wiring diagrams at the end of the book).
5   If there's battery voltage when there should be in Step 4, connect an ohmmeter from the blue/red wire's terminal in the harness side of the connector to a good ground/earth. With the engine cold, the resistance reading should be as listed in this Chapter's Specifications. If not, check the blue/red wire and the connector for breaks or bad connections.
6   If resistance is as specified, check the oil temperature sensor.

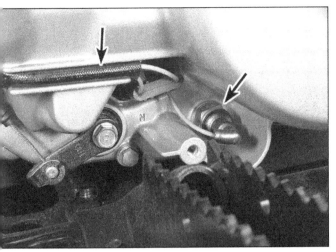

12.12a The oil temperature switch (right arrow) is mounted in the crankcase; the wiring harness (left arrow) fits in a specially cast crankcase groove

12.12b If you remove the switch wiring harness, don't forget to clip it into this retainer (arrow) on installation

14   Fill the crankcase with the recommended type and amount of oil (see Chapter 1) and check for leaks.

13   Ignition main (key) switch - check and replacement

Check

1   Follow the wiring harness from the ignition switch to the connector and unplug the connector.
2   Using an ohmmeter, check the continuity of the terminal pairs indicated in the wiring diagrams at the end of the book. Continuity should exist between the terminals connected by a solid line when the switch is in the indicated position.
3   If the switch fails any of the tests, replace it.

Replacement

Refer to illustration 13.4
4   The ignition switch is secured to the handlebar cover by two plastic prongs **(see illustration)**.
5   If you haven't already done so, unplug the switch electrical connector. Squeeze the prongs and lift the switch out of the handlebar cover.
6   Installation is the reverse of the removal procedure.

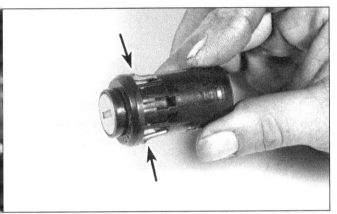

13.4 Squeeze the prongs (arrows) to detach the switch from the handlebar cover

Oil temperature sensor

Check

7   Remove the seat (see Chapter 7). Locate the three-pin connector in front of the air cleaner housing and unplug it.
8   With the engine cold, connect an ohmmeter between ground/earth and the blue wire's terminal in the switch side of the connector (it should indicate the same reading as in Step 5). Now warm up the engine to normal operating temperature. The resistance reading should now be considerably less (see this Chapter's Specifications).
9   If the sensor doesn't operate as described, replace it. If the sensor is good and you've already checked the wiring for breaks and bad connections, the most likely cause of the problem is a defective alarm unit.

Replacement

Refer to illustrations 12.11, 12.12a and 12.12b
10   Drain the engine oil (see Chapter 1).
11   Remove the switch cover from the right side of the engine **(see illustration)**.
12   Disconnect the electrical connector from the sensor and unscrew it from the engine **(see illustrations)**.
13   Wrap the threads of the sending unit with Teflon tape or apply a thin coat of sealant to them, then screw the unit into its hole, tightening it to the torque listed in this Chapter's Specifications. The remainder of installation is the reverse of removal.

14   Handlebar switches - check

1   Generally speaking, the switches are reliable and trouble-free. Most troubles, when they do occur, are caused by dirty or corroded contacts, but wear and breakage of internal parts is a possibility that should not be overlooked. If breakage does occur, the entire switch and related wiring harness will have to be replaced with a new one, since individual parts are not usually available.
2   The switches can be checked for continuity with an ohmmeter or a continuity test light. Always disconnect the battery negative cable, which will prevent the possibility of a short circuit, before making the checks.
3   Trace the wiring harness of the switch in question and unplug the electrical connectors.
4   Using the ohmmeter or test light, check for continuity between the terminals of the switch harness with the switch in the various positions. Refer to the continuity diagrams contained in the wiring diagrams at the end of the book. Continuity should exist between the terminals connected by a solid line when the switch is in the indicated position.

15.1 The handlebar switches and the choke lever are mounted in the left handlebar housing

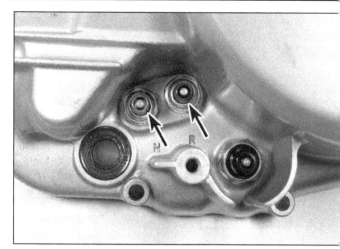

16.2 The neutral and reverse switches are located on the right side of the engine behind the switch cover (arrows); they're labeled N for Neutral and R for Reverse

5   If the continuity check indicates a problem, refer to Section 15, disassemble the switch and spray the switch contacts with electrical contact cleaner. If they are accessible, the contacts can be scraped clean with a knife or polished with crocus cloth. If switch components are damaged or broken, it will be obvious when the switch is disassembled.

15   Handlebar switches - removal and installation

Refer to illustration 15.1

1   The handlebar switches are composed of two halves that clamp around the bars. They are easily removed for cleaning or inspection by taking out the clamp screws and pulling the switch halves away from the handlebars **(see illustration)**.
2   To completely remove the switches, the electrical connectors in the wiring harness must be unplugged and the harness separated from the tie wraps and retainers.
3   When installing the switches, make sure the wiring harness is properly routed to avoid pinching or stretching the wires.

16   Neutral and reverse switches - check and replacement

Refer to illustrations 16.2 and 16.7

1   Remove the switch cover from the right side of the engine and disconnect the wiring connector from the reverse and neutral switches **(see illustration 12.11)**. **Warning:** *Label the reverse and neutral switch wires so there's no chance of mixing them up when you reconnect them. If this happens, the neutral light will come on when the transmission is actually in reverse. This may cause the vehicle to move backward when you don't expect it.*
2   Connect one lead of an ohmmeter to a good ground/earth and the other lead to the terminal post on the switch being tested **(see illustration)**.
3   When the transmission is in neutral, the ohmmeter should read 0 ohms between the neutral switch and ground/earth - in any other position, the meter should read infinite resistance.
4   When the transmission is in reverse, the ohmmeter should read 0 ohms between the reverse switch and ground/earth - in any other position, the meter should read infinite resistance.
5   Unscrew the neutral switch from the case. Connect the ohmmeter between the terminal post on the switch and the switch body. It should read 0 ohms when the switch is pressed in and infinite resistance when it's released.

16.7 Check the reverse/neutral switch rotor (arrow) for damage or a loose bolt

6   If the switch doesn't check out as described, replace it.
7   If the switch tests as defective on the vehicle, but tests okay when it's removed, remove the right side (clutch) cover from the crankcase (see Chapter 2). Check the switch rotor for damage or looseness **(see illustration)**.
8   Wrap the threads of the switch with Teflon tape or apply a thin coat of RTV sealant to them. Install the switch in the case and tighten it to the torque listed in this Chapter's Specifications.
9   Reconnect the switch wires, taking note of the Warning in Step 1.
10   The remainder of installation is the reverse of the removal steps.

17   Starter switch - check and replacement

The starter is switch is part of the switch assembly on the left handlebar. Refer to Section 14 for checking and replacement procedures.

18   Starter relay - check and replacement

Refer to illustration 18.3

1   Refer to Section 3 and remove the battery cover.
2   Disconnect the negative cable from the battery.

18.3 The relay terminals are located beneath plastic covers (A); the two-wire connector is mounted in the connector block (B); the tabs on the mounting bracket (C) engage the slots in the rubber holder (D)

3 Pull back the rubber covers from the terminal nuts, remove the nuts and disconnect the starter relay cables **(see illustration)**. Disconnect the remaining electrical connector from the connector block next to the starter relay.
4 Connect an ohmmeter between the terminals from which the nuts were removed. It should indicate infinite resistance.
5 Connect a 12-volt battery to the terminals that were disconnected from the connector block (the side that runs to the relay). The motorcycle's battery can be used if it's fully charged. The ohmmeter should now indicate zero ohms.
6 If the relay doesn't perform as described, pull its rubber mount off the metal bracket and pull the relay out of the mount.
7 Installation is the reverse of removal. Reconnect the negative battery cable after all the other electrical connections are made.

19 Starter diode - removal, check and installation

Refer to illustrations 19.2 and 19.3
1 Remove the front fender (see Chapter 3).
2 Remove the connector box cover bolt and open the box **(see illustration)**.

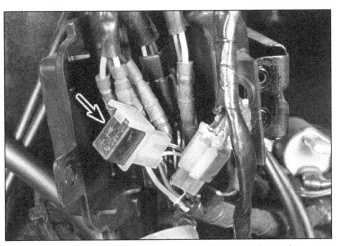

19.3 Open the cover and unplug the diode (arrow) from the connector

19.2 Remove the lower bolt (arrow) to open the diode cover

3 Unplug the diode from the wiring harness **(see illustration)**.
4 Connect an ohmmeter between the diode terminals, then switch the ohmmeter leads so they're connected to the opposite terminals. There should be continuity (little or no resistance) in one direction and infinite resistance in the other direction. If the diode doesn't check out okay, replace it.
5 Installation is the reverse of the removal steps.

20 Starter motor - removal and installation

Removal

Refer to illustrations 20.3 and 20.6
1 Disconnect the cable from the negative terminal of the battery.
2 Refer to Chapter 2 and remove the outermost starter reduction gear.
3 Pull back the rubber boot and remove the nut retaining the starter cable to the starter **(see illustration)**.
4 Remove the starter mounting bolts (one bolt secures a ground/earth wire).
5 Lift the outer end of the starter up a little bit and slide the starter out of the engine case. **Caution:** *Don't drop or strike the starter or its magnets may be demagnetized, which will ruin it.*

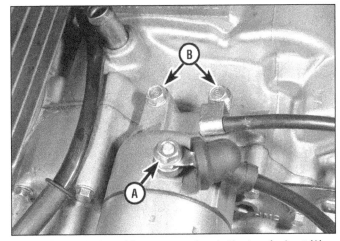

20.3 Pull back the rubber cover and undo the terminal nut (A), then remove the mounting bolts (B); the mounting bolt on the right also secures a ground/earth wire

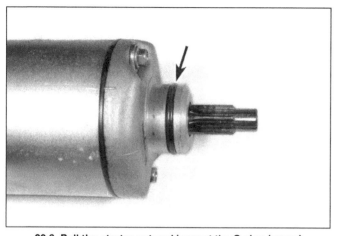

20.6 Pull the starter out and inspect the O-ring (arrow)

21.2 Make alignment marks between the housing and end covers before disassembly

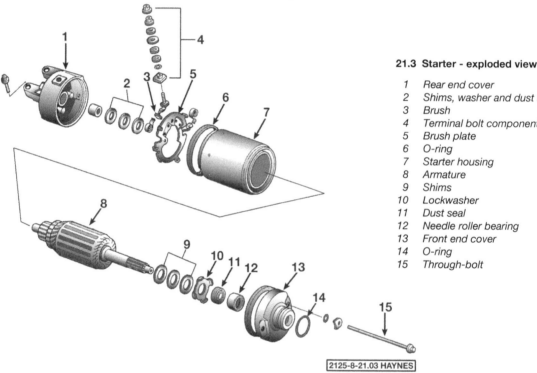

21.3 Starter - exploded view

1   *Rear end cover*
2   *Shims, washer and dust seal*
3   *Brush*
4   *Terminal bolt components*
5   *Brush plate*
6   *O-ring*
7   *Starter housing*
8   *Armature*
9   *Shims*
10   *Lockwasher*
11   *Dust seal*
12   *Needle roller bearing*
13   *Front end cover*
14   *O-ring*
15   *Through-bolt*

2125-8-21.03 HAYNES

6   Check the condition of the O-ring on the end of the starter and replace it if necessary **(see illustration)**.

Installation

7   Remove any corrosion or dirt from the mounting lugs on the starter and the mounting points on the crankcase.
8   Apply a little engine oil to the O-ring and install the starter by reversing the removal procedure.

21   Starter motor - disassembly, inspection and reassembly

1   Remove the starter motor (see Section 20).

Disassembly

Refer to illustrations 21.2 and 21.3
2   Make alignment marks between the housing and covers **(see illustration)**.

3   Unscrew the two long bolts, then remove the cover with its O-ring from the motor. Remove the shim(s) from the armature, noting their correct locations **(see illustration)**.
4   Remove the front cover with its O-ring from the motor. Remove the toothed washer from the cover and slide off the insulating washer and shim(s) from the front end of the armature, noting their locations.
5   Withdraw the armature from the housing.

Inspection

Refer to illustrations 21.7, 21.8, 21.9, 21.11a and 21.11b
Note: *Check carefully which components are available as replacements before starting overhaul procedures.*
6   Connect an ohmmeter between the terminal bolt and the insulated brush holder (or the indigo colored wire). There should be continuity (little or no resistance). When the ohmmeter is connected between the rear cover and the insulated brush holder (or the indigo colored wire), there should be no continuity (infinite resistance). Replace the brush holder if it doesn't test as described.
7   Unscrew the nut from the terminal bolt and remove the plain

21.7 Unscrew the nut and remove the washers from the terminal bolt, noting their correct installed order

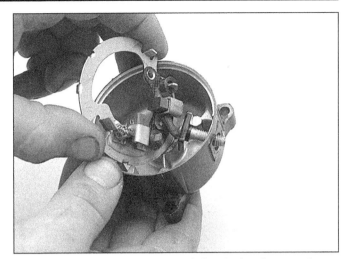

21.8 Lift the brush springs and slide the brushes out of their holders

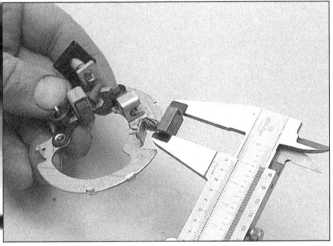

21.9 Measure the brush length and replace them if they're worn

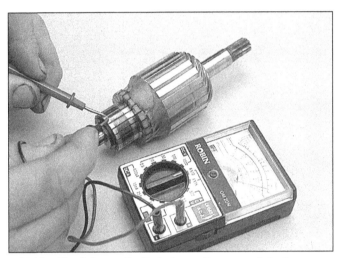

21.11a Check for continuity between the commutator bars . . .

washer, insulating washers and rubber ring, noting carefully how they're installed (see illustration). Withdraw the terminal bolt and brush assembly from the housing and recover the insulator.

8   Lift the brush springs and slide the brushes out of their holders (see illustration).

9   The parts of the starter motor that most likely will require attention are the brushes. If one brush must be replaced, replace both of them. The brushes are replaced together with the terminal bolt and the brush plate. Brushes must be replaced if they are worn excessively, cracked, chipped, or otherwise damaged. Measure the length of the brushes and compare the results to the brush length listed in this Chapter's Specifications (see illustration). If either of the brushes is worn beyond the specified limits, replace them both.

10   Inspect the commutator for scoring, scratches and discoloration. The commutator can be cleaned and polished with fine emery paper, but do not use sandpaper and do not remove copper from the commutator. After cleaning, clean out the grooves and wipe away any residue with a cloth soaked in an electrical system cleaner or denatured alcohol.

11   Using an ohmmeter or a continuity test light, check for continuity between the commutator bars (see illustration). Continuity should exist between each bar and all of the others. Also, check for continuity between the commutator bars and the armature shaft (see illustration). There should be no continuity between the commutator and the shaft. If the checks indicate otherwise, the armature is defective.

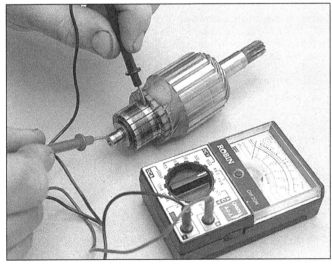

21.11b . . . and for no continuity between each commutator bar and the armature shaft

21.16 Install the brush plate assembly and terminal bolt

21.17 Install the terminal bolt nut and tighten it securely

21.19a Install the shims on the armature, then insert the armature . . .

21.19b . . . and locate the brushes on the commutator

21.20 Fit the lockwasher to the front cover so its teeth engage the cover ribs

12 Check the dust seal in the front cover for wear or damage. Check the needle roller bearing in the front cover for roughness, looseness or loss of lubricant. Check with a motorcycle shop or Honda dealer to see if the bearing can be replaced separately; if this isn't possible, replace the starter motor.

13 Inspect the bushing in the rear cover. Replace the starter motor if the bushing is worn or damaged.

14 Check the starter pinion for worn, chipped or broken teeth. If the gear is damaged or worn, replace the starter motor. Inspect the insulating washers for signs of damage and replace if necessary.

Reassembly

Refer to illustrations 21.16, 21.17, 21.19a, 21.19b, 21.20, 21.21, 21.22 and 21.23

15 Lift the brush springs and slide the brushes back into position in their holders.

16 Fit the insulator to the housing and install the brush plate. Insert the terminal bolt through the brush plate and housing **(see illustration)**.

17 Slide the rubber ring and small insulating washer onto the bolt, followed by the large insulating washers and the plain washer. Fit the nut to the terminal bolt and tighten it securely **(see illustration)**.

18 Locate the brush assembly in the housing, making sure its tab is correctly located in the housing slot.

19 Fit the shims to the armature shaft **(see illustration)** and insert the armature in the housing, locating the brushes to the commutator bars. Check that each brush is securely pressed against the commutator by its spring and is free to move easily in its holder **(see illustration)**.

21.21 Install the shims and washer on the armature, making sure they're in the correct order

20 Fit the lockwasher to the front cover so that its teeth are correctly located with the cover ribs **(see illustration)**. Apply a smear of grease to the cover dust seal lip.

21 Slide the shim(s) onto the front end of the armature shaft, then fit the insulating washer **(see illustration)**. Fit the sealing ring to the housing and carefully slide the front cover into position, aligning the

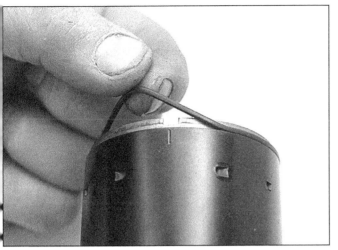

21.22 Fit the O-ring . . .

21.23 . . . and install the rear cover, aligning its groove
with the brush plate outer tab

22.1 Remove the reduction gear cover bolts (the screws
needn't be removed) . . .

22.2 . . . and pull the cover off; the gears may come with it

marks made on removal.

22   Ensure the brush plate inner tab is correctly located in the housing slot and fit the O-ring to the housing **(see illustration)**.

23   Align the rear cover groove with the brush plate outer tab and install the cover **(see illustration)**.

24   Check that the marks made on disassembly are correctly aligned, then fit the long bolts and tighten them securely **(see illustration 21.3)**.

25   Install the starter as described in Section 20.

22   Starter reduction gears - removal, inspection, bearing replacement and installation

Removal

Refer to illustrations 22.1, 22.2, 22.3a, 22.3b and 22.3c

1   Unbolt the reduction gear cover and take it off the engine **(see illustration)**.

2   The gears may come off with the cover **(see illustration)**. If not, remove them separately.

3   If the gears came off with the cover, remove them **(see illustrations)**.

22.3a The assembled reduction gears look like this . . .

22.3b Pull out the shaft and snap-ring . . .

22.3c . . . then pull out the remaining gears and washer

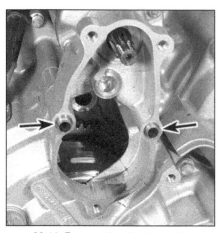

22.11 Be sure the dowels are in place (arrows)

23.2 Hold the alternator rotor and try to turn the starter driven gear; it should turn smoothly clockwise, but not at all counterclockwise (anti-clockwise)

23.3 Lift the driven gear off the rotor

Inspection

4   Check the gears for worn or broken teeth. Check the shafts and the friction surface on the gears for wear or damage and replace any parts that show defects.

5   Turn the bearing with a finger. If it's rough, loose or noisy, replace it.

Bearing replacement

6   Remove the plastic protector from the aluminum reduction gear cover.

7   Heat the cover in an oven, then hold it with the bearing facing down and tap around the bearing with a soft faced mallet until the bearing drops out.

8   Tap a new bearing in with a socket or bearing driver that bears against the bearing outer race.

9   Install the plastic protector on the reduction gear cover.

Installation

Refer to illustration 22.11

10   Remove all old gasket and sealant from the cover and crankcase.

11   Install a new gasket and make sure the dowels are in position **(see illustration)**.

12   Assemble the gears in the cover **(see illustration 22.2)**.

13   Install the cover and gears on the crankcase. Install the bolts and tighten them evenly in a criss-cross pattern to the torque listed in this Chapter's Specifications.

---

23 Starter clutch - removal, inspection and installation

Refer to illustrations 23.2, 23.3, 23.4a and 23.4b

1   Remove the alternator cover and rotor (see Section 26). The starter clutch is mounted on the back of the alternator rotor.

2   Hold the alternator rotor with one hand and try to rotate the starter driven gear with the other hand **(see illustration)**. It should rotate clockwise smoothly, but not rotate counterclockwise (anti-clockwise) at all.

3   If the gear rotates both ways or neither way, or if its movement is rough, remove it from the alternator rotor **(see illustration)**.

4   Remove the Torx bolts and lift off the outer clutch **(see illustrations)**. Inspect the outer clutch and one-way clutch for wear and damage and replace them if any problems are found.

5   Installation is the reverse of the removal steps. Before you tighten the Torx bolts, turn the starter driven gear (see Step 2). It should rotate clockwise only; if it rotates counterclockwise (anti-clockwise), the one-way clutch is installed backwards. Apply non-permanent thread locking agent to the threads of the Torx bolts and tighten them to the torque listed in this Chapter's Specifications.

23.4a Remove the six Torx bolts . . .

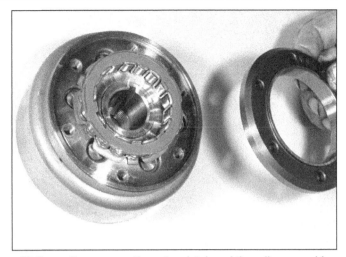

23.4b . . . then remove the outer clutch and the roller assembly

24 Charging system testing - general information and precautions

1   If the performance of the charging system is suspect, the system as a whole should be checked first, followed by testing of the individual components (the alternator and the regulator/rectifier). **Note:** *Before beginning the checks, make sure the battery is fully charged and that all system connections are clean and tight.*
2   Checking the output of the charging system and the performance of the various components within the charging system requires the use of a voltmeter, ammeter and ohmmeter or the equivalent multimeter.
3   When making the checks, follow the procedures carefully to prevent incorrect connections or short circuits, as irreparable damage to electrical system components may result if short circuits occur.
4   If the necessary test equipment is not available, it is recommended that charging system tests be left to a dealer service department or a reputable motorcycle repair shop.

25 Charging system - leakage and output test

1   If a charging system problem is suspected, perform the following checks. Start by removing the battery cover (see Section 3).

Leakage test

Refer to illustration 25.3
2   Turn the ignition switch Off and disconnect the cable from the battery negative terminal.
3   Set the multimeter to the mA (milliamps) function and connect its negative probe to the battery negative terminal, and the positive probe to the disconnected negative cable **(see illustration)**. Compare the reading to the leakage limit listed in this Chapter's Specifications.
4   If the reading exceeds the specifications there is probably a short circuit in the wiring. Thoroughly check the wiring between the various components (see the wiring diagrams at the end of the book).
5   If the reading is below the specified amount, the leakage rate is satisfactory. Disconnect the meter and connect the negative cable to the battery, tightening it securely. Check the alternator output as described below.

Output test

6   Start the engine and let it warm up to normal operating temperature.
7   With the engine idling, attach the positive (red) voltmeter lead to the positive (+) battery terminal and the negative (black) lead to the

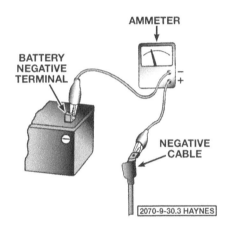

25.3 Checking the charging system leakage rate with an ammeter

battery negative (-) terminal. The voltmeter selector switch (if equipped) must be in the 0-20 DC volt range.
8   Slowly increase the engine speed to 5000 rpm and compare the voltmeter reading to the value listed in this Chapter's Specifications.
9   If the output is as specified, the alternator is functioning properly.
10   Low voltage output may be the result of damaged windings in the alternator stator coils or wiring problems. Make sure all electrical connections are clean and tight, then refer to the following Sections to check the alternator stator coils and the regulator/rectifier.
11   High voltage output (above the specified range) indicates a defective voltage regulator/rectifier. Refer to Section 27 for regulator testing and replacement procedures.

26 Alternator stator coils and rotor - check and replacement

Stator coil check

1   Locate and disconnect the alternator coil connector on the left side of the vehicle frame **(see illustration 4.1 in Chapter 5)**.
2   Connect an ohmmeter between the terminals in the side of the connector that runs back to the stator coils on the left side of the engine. If the readings are much outside the value listed in this Chapter's Specifications, replace the stator coils as described below.
3   Connect the ohmmeter between a good ground/earth on the

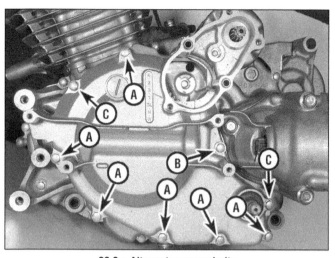

26.9a Alternator cover bolts

A   Bolts   C   Dowel locations
B   Bolt with copper washer

26.9b Pull the cover and wiring harness off the crankcase; the rotor magnets may provide some resistance, but DO NOT force or pry the cover off

vehicle and each of the connector terminals in turn. The meter should indicate infinite resistance (no continuity). If not, replace the stator coils.

Stator coil replacement

Refer to illustrations 26.9a, 26.9b, 26.10, 26.11 and 26.12

4   Drain the engine oil (see Chapter 1).
5   If you're working on a 4WD model, remove the engine skid plate and the front drive side shaft (see Chapter 5).
6   Remove the left footpeg (see Chapter 7).
7   Remove the shift pedal (see Chapter 2).
8   Remove the starter reduction gears (see Section 22).
9   Remove the alternator cover bolts (and the sealing washer that fits on one of the bolts) and carefully separate the cover from the engine **(see illustrations)**. The rotor magnets may create some resistance, so pull firmly if necessary. Don't force the cover off, though, and don't pry against the gasket surfaces; if the cover won't come off, make sure all fasteners have been removed.
10   Remove the wire clamp bolt and peel the sealant away from the wiring harness grommet **(see illustration)**.
11   Remove the stator coil Allen bolts and take the stator coils out of the cover **(see illustration)**.
12   Installation is the reverse of the removal steps, with the following additions:

a) *Tighten the stator coil Allen bolts to the torque listed in this Chapter's Specifications.*
b) *Remove all old gasket material from the alternator cover and crankcase. Use a new gasket on the alternator cover and smear a film of sealant across the wiring harness grommet* **(see illustration 26.10)**.
c) *Make sure the cover dowels are in position* **(see illustration)**.
d) *Be sure to install the sealing washer on the cover bolt with the arrowhead cast next to it* **(see illustration 26.9a)**. *Tighten the cover bolts evenly, in a criss-cross pattern, to the torque listed in this Chapter's Specifications.*

Rotor replacement

Removal

Refer to illustrations 26.14, 26.15, 26.16, 26.17 and 26.18

Note: *To remove the alternator rotor, the special Honda puller (part no. 07733-0020001 or 07933-3950000) or an aftermarket equivalent will be required. Don't try to remove the rotor without the proper puller, as it's almost sure to be damaged. Pullers are readily available from motorcycle dealers and aftermarket tool suppliers.*

13   Remove the alternator cover (Steps 4 through 9).
14   Hold the alternator rotor with a strap wrench. If you don't have one and the engine is in the frame, the rotor can be locked by placing the transmission in gear and holding the rear brake on. Unscrew the

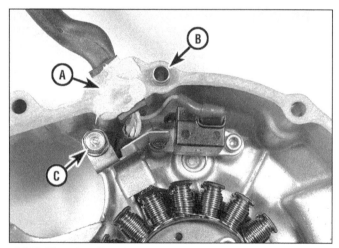

26.10 There's a film of sealant across the grommet location (A); the dowels (B) may stay in the engine or come off with the cover; remove the bolt (C) to detach the wiring harness

26.11 The stator coils are secured by three Allen bolts (arrows)

26.12 **Make sure the dowels are in position (arrows)**

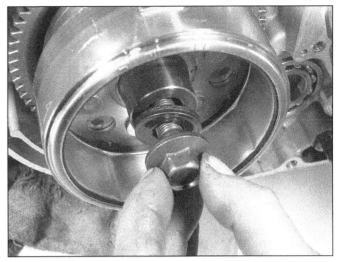

26.14 **Remove the rotor bolt and washer**

rotor bolt and remove the washer **(see illustration)**.
15   Thread an alternator puller into the center of the rotor and use it to remove the rotor **(see illustration)**. If the rotor doesn't come off easily, tap sharply on the end of the puller to release the rotor's grip on the tapered crankshaft end.
16   Pull the rotor off, together with the starter clutch and starter driven gear **(see illustration)**.
17   Remove the needle roller bearing and washer from the crankshaft **(see illustration)**.
18   Check the Woodruff key **(see illustration)**; if it's not secure in its slot, pull it out and set it aside for safekeeping. A convenient method is to stick the Woodruff key to the magnets inside the rotor, but be certain not to forget it's there, as serious damage to the rotor and stator coils will occur if the engine is run with anything stuck to the magnets.

Installation

Refer to illustration 26.23
19   Degrease the center of the rotor and the end of the crankshaft.
20   Make sure the Woodruff key is positioned securely in its slot **(see illustration 26.18)**.
21   Align the rotor slot with the Woodruff key. Place the rotor, together with the starter clutch and starter driven gear, on the crankshaft.

26.15 **Remove the rotor with a puller designed for the purpose**

26.16 **Pull the rotor off together with the starter clutch and driven gear**

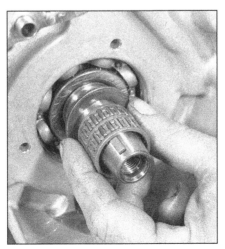

26.17 **Remove the needle roller bearing and washer**

26.18 **Make sure the Woodruff key (arrow) is securely in its slot before installing the rotor**

26.23 Be sure there aren't any small metal objects stuck to the rotor magnets; an inconspicuous item like this Woodruff key (arrow) can ruin the rotor and stator if the engine is run

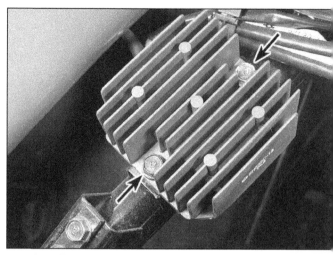

27.1 The regulator/rectifier unit is secured to the frame by two bolts (arrows)

22   Install the washer and rotor bolt. Hold the rotor from turning with one of the methods described in Step 14 and tighten the bolt to the torque listed in this Chapter's Specifications.
23   Take a look to make sure there isn't anything stuck to the inside of the rotor **(see illustration)**.

---

27 Regulator/rectifier - check and replacement

Check

Refer to illustration 27.1
1   The regulator/rectifier is mounted on the left side of the vehicle frame under the rear fender **(see illustration)**.
2   Follow the wiring harnesses from the regulator/rectifier to the connectors and unplug the connectors. The connectors are located alongside the air cleaner housing and in the battery compartment. They can be identified by their wire colors (refer to the Wiring diagrams).
3   Connect a voltmeter between the red wire terminal and the green wire terminal in the harness side of the connector. It should indicate battery voltage (approximately 13 volts).
4   Connect the voltmeter between the black and green wire terminals in the harness side of the connector. With the ignition On, the meter should indicate battery voltage.
5   If the voltage readings aren't as described, check the wiring for breaks or poor connections.
6   Make the following checks with an ohmmeter. Be sure the ohmmeter battery is in good condition, and don't allow your fingers to touch the ohmmeter probes or inaccurate readings may occur. Honda recommends the following brands and models of ohmmeter:
  a) *Kowa digital 07411-0020000*
  b) *Kowa digital KS-AHM-32-003 (US only)*
  c) *Kowa analog TH-5H*
  d) *Sanwa analog 07308-0020001*
Other ohmmeters may produce different readings.
7   Connect the negative probe to the red wire terminal in the side of the connector that runs back to the regulator/rectifier unit. Connect the positive probe to the following connectors in turn and note the readings:
  a) *To green: infinite resistance*
  b) *To black: infinite resistance*
  c) *To yellow: infinite resistance*
8   Connect the ohmmeter negative probe to the green wire terminal, then connect the positive probe to the following connectors in turn and note the readings:
  a) *To red: 1 to 20 ohms*
  b) *To black: 1 to 20 ohms*
  c) *To yellow: 0.5 to 10 ohms*
9   Connect the ohmmeter negative probe to the black wire terminal, then connect the positive probe to the following connectors in turn and note the readings:
  a) *To red: 20 to 100 ohms*
  b) *To green: 10 to 50 ohms*
  c) *To yellow: 15 to 80 ohms*
10   Connect the ohmmeter negative probe to the yellow wire terminal, then connect the positive probe to the following connectors in turn and note the readings:
  a) *To red: 0.5 to 10 ohms*
  b) *To green: infinite resistance*
  c) *To black: infinite resistance*
11   If the regulator/rectifier doesn't test as described, it may be defective. It's a good idea to have it tested by a Honda dealer or substitute a known good unit before condemning it as trash.

---

28 Wiring diagrams

Prior to troubleshooting a circuit, check the fuses to make sure they're in good condition. Make sure the battery is fully charged and check the cable connections.
When checking a circuit, make sure all connectors are clean, with no broken or loose terminals or wires. When unplugging a connector, don't pull on the wires - pull only on the connector housings themselves.

Wiring diagram color key

Bl - black	Lg - light green
Br - brown	O - orange
Bu - blue	P - pink
G - green	R - red
Gr - gray	W - white
Lb - light blue	Y - yellow

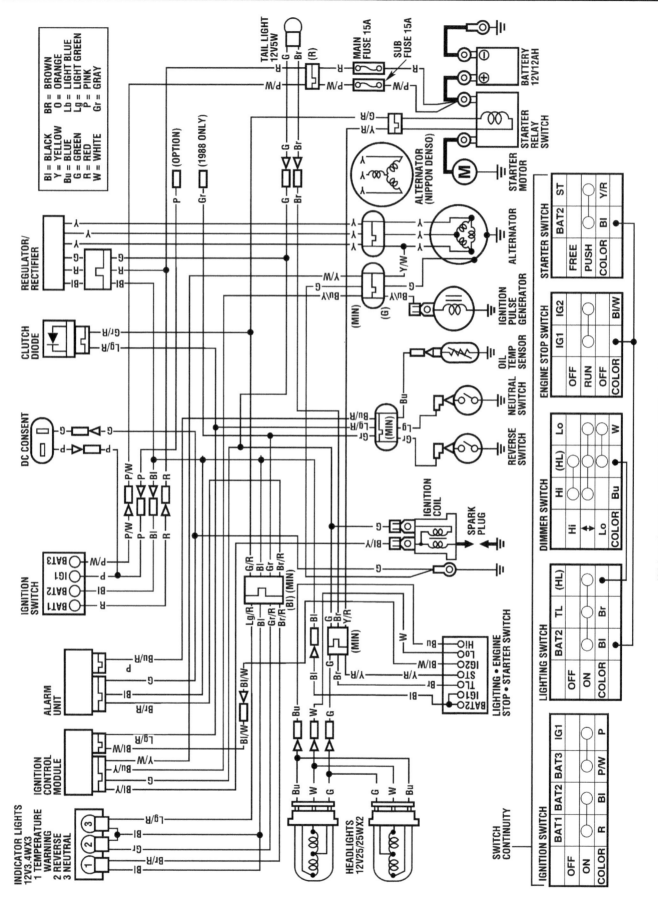

Wiring diagram - 1988 through 1992 models

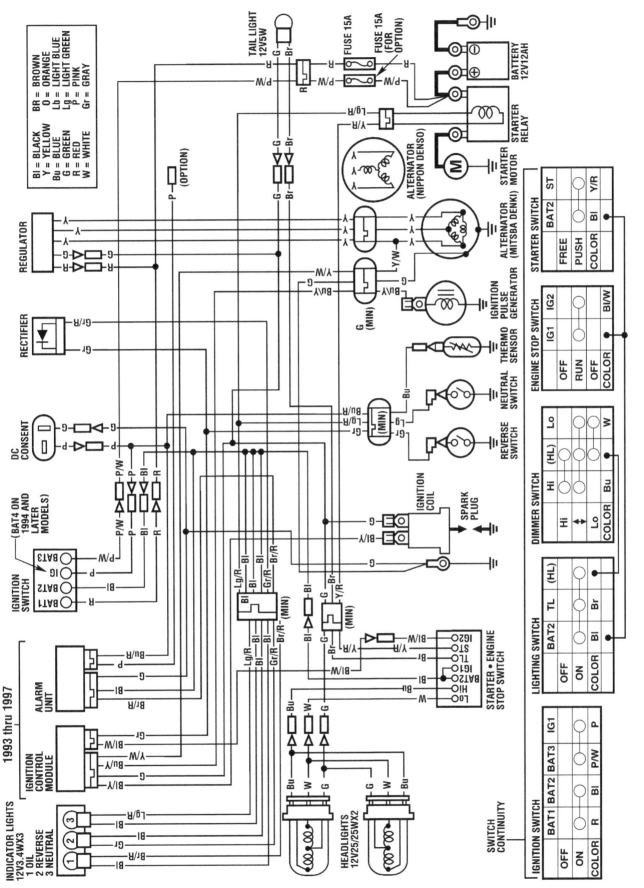

Wiring diagram - 1993 and later models

Conversion factors

Length (distance)

Inches (in)	X 25.4	= Millimetres (mm)	X 0.0394	= Inches (in)
Feet (ft)	X 0.305	= Metres (m)	X 3.281	= Feet (ft)
Miles	X 1.609	= Kilometres (km)	X 0.621	= Miles

Volume (capacity)

Cubic inches (cu in; in³)	X 16.387	= Cubic centimetres (cc; cm³)	X 0.061	= Cubic inches (cu in; in³)
Imperial pints (Imp pt)	X 0.568	= Litres (l)	X 1.76	= Imperial pints (Imp pt)
Imperial quarts (Imp qt)	X 1.137	= Litres (l)	X 0.88	= Imperial quarts (Imp qt)
Imperial quarts (Imp qt)	X 1.201	= US quarts (US qt)	X 0.833	= Imperial quarts (Imp qt)
US quarts (US qt)	X 0.946	= Litres (l)	X 1.057	= US quarts (US qt)
Imperial gallons (Imp gal)	X 4.546	= Litres (l)	X 0.22	= Imperial gallons (Imp gal)
Imperial gallons (Imp gal)	X 1.201	= US gallons (US gal)	X 0.833	= Imperial gallons (Imp gal)
US gallons (US gal)	X 3.785	= Litres (l)	X 0.264	= US gallons (US gal)

Mass (weight)

Ounces (oz)	X 28.35	= Grams (g)	X 0.035	= Ounces (oz)
Pounds (lb)	X 0.454	= Kilograms (kg)	X 2.205	= Pounds (lb)

Force

Ounces-force (ozf; oz)	X 0.278	= Newtons (N)	X 3.6	= Ounces-force (ozf; oz)
Pounds-force (lbf; lb)	X 4.448	= Newtons (N)	X 0.225	= Pounds-force (lbf; lb)
Newtons (N)	X 0.1	= Kilograms-force (kgf; kg)	X 9.81	= Newtons (N)

Pressure

Pounds-force per square inch (psi; lbf/in²; lb/in²)	X 0.070	= Kilograms-force per square centimetre (kgf/cm²; kg/cm²)	X 14.223	= Pounds-force per square inch (psi; lbf/in²; lb/in²)
Pounds-force per square inch (psi; lbf/in²; lb/in²)	X 0.068	= Atmospheres (atm)	X 14.696	= Pounds-force per square inch (psi; lbf/in²; lb/in²)
Pounds-force per square inch (psi; lbf/in²; lb/in²)	X 0.069	= Bars	X 14.5	= Pounds-force per square inch (psi; lbf/in²; lb/in²)
Pounds-force per square inch (psi; lbf/in²; lb/in²)	X 6.895	= Kilopascals (kPa)	X 0.145	= Pounds-force per square inch (psi; lbf/in²; lb/in²)
Kilopascals (kPa)	X 0.01	= Kilograms-force per square centimetre (kgf/cm²; kg/cm²)	X 98.1	= Kilopascals (kPa)
Millibar (mbar)	X 100	= Pascals (Pa)	X 0.01	= Millibar (mbar)
Millibar (mbar)	X 0.0145	= Pounds-force per square inch (psi; lbf/in²; lb/in²)	X 68.947	= Millibar (mbar)
Millibar (mbar)	X 0.75	= Millimetres of mercury (mmHg)	X 1.333	= Millibar (mbar)
Millibar (mbar)	X 0.401	= Inches of water (inH₂O)	X 2.491	= Millibar (mbar)
Millimetres of mercury (mmHg)	X 0.535	= Inches of water (inH₂O)	X 1.868	= Millimetres of mercury (mmHg)
Inches of water (inH₂O)	X 0.036	= Pounds-force per square inch (psi; lbf/in²; lb/in²)	X 27.68	= Inches of water (inH₂O)

Torque (moment of force)

Pounds-force inches (lbf in; lb in)	X 1.152	= Kilograms-force centimetre (kgf cm; kg cm)	X 0.868	= Pounds-force inches (lbf in; lb in)
Pounds-force inches (lbf in; lb in)	X 0.113	= Newton metres (Nm)	X 8.85	= Pounds-force inches (lbf in; lb in)
Pounds-force inches (lbf in; lb in)	X 0.083	= Pounds-force feet (lbf ft; lb ft)	X 12	= Pounds-force inches (lbf in; lb in)
Pounds-force feet (lbf ft; lb ft)	X 0.138	= Kilograms-force metres (kgf m; kg m)	X 7.233	= Pounds-force feet (lbf ft; lb ft)
Pounds-force feet (lbf ft; lb ft)	X 1.356	= Newton metres (Nm)	X 0.738	= Pounds-force feet (lbf ft; lb ft)
Newton metres (Nm)	X 0.102	= Kilograms-force metres (kgf m; kg m)	X 9.804	= Newton metres (Nm)

Power

Horsepower (hp)	X 745.7	= Watts (W)	X 0.0013	= Horsepower (hp)

Velocity (speed)

Miles per hour (miles/hr; mph)	X 1.609	= Kilometres per hour (km/hr; kph)	X 0.621	= Miles per hour (miles/hr; mph)

Fuel consumption*

Miles per gallon, Imperial (mpg)	X 0.354	= Kilometres per litre (km/l)	X 2.825	= Miles per gallon, Imperial (mpg)
Miles per gallon, US (mpg)	X 0.425	= Kilometres per litre (km/l)	X 2.352	= Miles per gallon, US (mpg)

Temperature

Degrees Fahrenheit = (°C x 1.8) + 32 Degrees Celsius (Degrees Centigrade; °C) = (°F - 32) x 0.56

It is common practice to convert from miles per gallon (mpg) to litres/100 kilometres (l/100km), where mpg (Imperial) x l/100 km = 282 and mpg (US) x l/100 km = 235

Index

A

About this manual, 0-5
Air cleaner housing, removal and installation, 3-9
Air filter element, cleaning, 1-11
Alternator stator coils and rotor, check and replacement, 8-15
Axle housing, removal and installation, 5-19

B

Battery
    charging, 8-3
    check, 1-6
    inspection and maintenance, 8-2
Bodywork and frame, 7-1 through 7-4
Brakes, wheels and tires, 6-1 through 6-16
    drum, removal, inspection and installation
        rear, 6-9
        front, 6-2
    front wheel cylinders and 2WD adjusters, removal, overhaul
        and installation, 6-5
    general information, 6-2
    hoses and lines, inspection and replacement, 6-9
    lever and pedal freeplay, check and adjustment, 1-7
    light
        bulbs, replacement, 8-5
        switches, check and replacement, 8-5
    master cylinder (front), removal, overhaul and installation, 6-7

    panel, removal, inspection and installation
        front, 6-7
        rear, 6-11
    pedal, lever and cables, removal and installation, 6-13
    shoes, removal, inspection and installation
        front, 6-4
        rear, 6-10
    system bleeding, 6-8
    system, general check, 1-6
    waterproof seal replacement, 6-3

C

Cam chain tensioner, removal and installation, 2-11
Camshaft and sprocket, removal, inspection and installation, 2-11
Carburetor
    disassembly, cleaning and inspection, 3-5
    overhaul, general information, 3-4
    reassembly and float height check, 3-8
    removal and installation, 3-4
Cautions, Notes and Warnings, 0-5
Charging system
    leakage and output test, 8-15
    testing, general information and precautions, 8-15
Chemicals and lubricants, 0-16
Choke
    cable, removal and installation, 3-10
    operation check, 1-9

Clutch
   check and freeplay adjustment, 1-9
   general information, 2-6
   removal, inspection and installation
      centrifugal clutch, 2-22
      change clutch, 2-25
Crankcase
   components, inspection and servicing, 2-39
   disassembly and reassembly, 2-38
**Crankshaft and balancer, removal, inspection
   and installation, 2-44**
Cylinder compression, check, 1-14
Cylinder head
   and valves, disassembly, inspection and reassembly, 2-14
   cover and rocker arms, removal, inspection and installation, 2-9
   removal and installation, 2-13
Cylinder, removal, inspection and installation, 2-17

D

Diagnosis, 0-17
**Differential and driveshaft, front (4WD), removal, inspection and
   installation, 5-13**
Differential oil, change, 1-10
Driveaxle, front (4WD)
   boot replacement and CV joint overhaul, 5-11
   removal and installation, 5-11
Driveshaft, rear, removal, inspection and installation, 5-23

E

Electrical system, 8-1 through 8-18
   general information, 8-2
   troubleshooting, 8-2
Engine
   cam chain tensioner, removal and installation, 2-11
   camshaft and sprocket, removal, inspection and installation, 2-11
   crankcase
      components, inspection and servicing, 2-39
      disassembly and reassembly, 2-38
   crankshaft and balancer, removal, inspection and installation, 2-44
   cylinder head
      and valves, disassembly, inspection and reassembly, 2-14
      cover and rocker arms, removal, inspection and installation, 2-9
      removal and installation, 2-13
   cylinder, removal, inspection and installation, 2-17
   disassembly and reassembly, general information, 2-8
   external shift mechanism, removal, inspection and installation, 2-35
   general information, 2-6
   initial start-up after overhaul, 2-46
   kickstarter, removal, inspection and installation, 2-33
   oil/filter, change, 1-10
   oil pump, removal, inspection and installation, 2-31
   operations possible with the engine in the frame, 2-6
   operations requiring engine removal, 2-7
   piston rings, installation, 2-21
   piston, removal, inspection and installation, 2-19
   primary drive gear, removal, inspection and installation, 2-33
   recommended break-in procedure, 2-47
   removal and installation, 2-7
   repair, major, general note, 2-7
   valves/valve seats/valve guides, servicing, 2-14
Engine, clutch and transmission, 2-1 through 2-48
Exhaust system
   inspection and spark arrester cleaning, 1-13
   removal and installation, 3-11

External oil pipe, removal and installation, 2-22
**External shift mechanism, removal, inspection
   and installation, 2-35**

F

Fasteners, check, 1-17
Fault finding, 0-17
Filter cleaning
   air,, 1-11
   fuel, 1-12
Fluid levels, check, 1-4
Footpegs, removal and installation, 7-4
Frame, general information, inspection and repair, 7-4
Frame, suspension and final drive, 5-1 through 5-24
Front carrier and fender, removal and installation, 7-1
**Front differential and driveshaft (4WD), removal, inspection
   and installation, 5-13**
**Front drive side shaft (4WD models) removal, inspection
   and installation, 5-14**
Front driveaxle (4WD)
   boot replacement and CV joint overhaul, 5-11
   removal and installation, 5-11
Fuel and exhaust systems, 3-1 through 3-12
   carburetor
      disassembly, cleaning and inspection, 3-5
      overhaul, general information, 3-4
      reassembly and float height check, 3-8
      removal and installation, 3-4
   choke cable, removal and installation, 3-10
   general information, 3-2
   idle fuel/air mixture adjustment, 3-3
   system check and filter cleaning, 1-12
   tank
      cleaning and repair, 3-3
      removal and installation, 3-2
   throttle cable and housing, removal, installation and adjustment, 3-9
Fuses, check and replacement, 8-3

G

General specifications, 0-7

H

Handlebars
   switches
      check, 8-7
      removal and installation, 8-8
   removal, inspection and installation, 5-3
Headlight
   aim, check and adjustment, 8-5
   bulb, replacement, 8-4

I

Identification numbers, 0-6
Idle fuel/air mixture adjustment, 3-3
Idle speed, check and adjustment, 1-16

Ignition system, 4-1 through 4-4
    coil, check, removal and installation, 4-2
    control module (ICM), harness check, removal and installation, 4-4
    main (key) switch, check and replacement, 8-7
    pulse generator, check, removal and installation, 4-3
    system
        check, 4-2
        general information, 4-2
    timing, general information and check, 4-4
Indicator bulbs, replacement, 8-6
Introduction to the Honda TRX300, 0-5

K

Kickstarter, removal, inspection and installation, 2-33

L

Lighting system, check, 8-4
Lubricants and chemicals, 0-16
Lubrication, general, 1-9

M

Maintenance
    schedule, 1-3
    techniques, 0-9

N

Neutral and reverse switches, check and replacement, 8-8
Notes, Cautions and Warnings, 0-5

O

Oil
    and filter, change, 1-10
    pipe and pump, removal, inspection and installation, 2-31
    pipe, external, removal and installation, 2-22
    temperature warning system, check and component
        replacement, 8-6

P

Piston rings, installation, 2-21
Piston, removal, inspection and installation, 2-19
Primary drive gear, removal, inspection and installation, 2-33
Pulse generator, check, removal and installation, 4-3

R

Rear axle, removal, inspection and installation, 5-17
Rear carrier and fender, removal and installation, 7-2
Rear driveshaft, removal, inspection and installation, 5-23
Rear final drive unit, removal, inspection and installation, 5-20

Regulator/rectifier, check and replacement, 8-18
Reverse lock mechanism, cable replacement, removal, inspection
    and installation, 2-29
Reverse lock system, check and adjustment, 1-8
Right side cover, removal and installation, 7-2
Routine maintenance, 1-1 through 1-18

S

Safe repair practices, 0-15
Seat, removal and installation, 7-1
Shift mechanism, external, removal, inspection
    and installation, 2-35
Shock absorbers
    disassembly, inspection and reassembly, 5-6
    removal and installation, 5-6
Side cover, right, removal and installation, 7-2
Spark arrester cleaning, 1-13
Spark plug, replacement, 1-13
Specifications, general, 0-7
Starter
    clutch, removal, inspection and installation, 8-14
    diode, removal, check and installation, 8-9
    motor
        disassembly, inspection and reassembly, 8-10
        removal and installation, 8-9
    reduction gears, removal, inspection, bearing replacement and
        installation, 8-13
    relay, check and replacement, 8-8
    switch, check and replacement, 8-8
Steering
    knuckle
        bearing and lower balljoint replacement, 5-10
        removal, inspection, and installation, 5-8
    shaft, removal, inspection, bearing replacement and installation, 5-5
    system, inspection and toe-in adjustment, 1-17
Suspension
    arms, removal, inspection, balljoint replacement
        and installation, 5-10
    check, 1-17
Swingarm
    bearings
        check, 5-21
        replacement, 5-23
    removal and installation, 5-21

T

Tail light and brake light bulbs, replacement, 8-5
Throttle
    cable and housing, removal, installation and adjustment, 3-9
    operation/grip freeplay, check and adjustment, 1-9
Tie-rods, removal, inspection and installation, 5-8
Tires
    and wheels, general check, 1-8
    general information, 6-15
Tools, 0-10
Trailer hitch, removal and installation, 7-4
Transfer case (4WD models)
    disassembly, inspection and reassembly, 5-16
    oil change, 1-10
    removal and installation, 5-15
Transmission
    general information, 2-6
    output gear and countershaft, removal, inspection and
        installation, 2-44
    shafts and shift drum, removal, inspection and installation, 2-40

Troubleshooting, 0-17
Tune-up and routine maintenance, 1-1 through 1-18

V

Valve clearances, check and adjustment, 1-15
Valves/valve seats/valve guides, servicing, 2-14

W

Warnings, Cautions and Notes, 0-5
Wheel bearing replacement (2WD), front, 6-3
Wheel hubs, rear, removal and installation, 6-15
Wheels, inspection, removal and installation, 6-14
Working facilities, 0-14